How We Think

HOW WE THINK

Digital Media and Contemporary Technogenesis

N. KATHERINE HAYLES

The University of Chicago Press
Chicago and London

N. Katherine Hayles is professor of literature at Duke University. Her books include *How We Became Posthuman: Virtual Bodies in Cybernetics, Literature, and Informatics* and *Writing Machines.*

The University of Chicago Press, Chicago 60637
The University of Chicago Press, Ltd., London

Printed in the United States of America

21 20 19 18 17 16 15 14 13 12   1 2 3 4 5

ISBN-13: 978-0-226-32140-0 (cloth)
ISBN-13: 978-0-226-32142-4 (paper)
ISBN-10: 0-226-32140-1 (cloth)
ISBN-10: 0-226-32142-8 (paper)

Library of Congress Cataloging-in-Publication Data

Hayles, N. Katherine.
How we think : digital media and contemporary technogenesis / N. Katherine Hayles.
p. cm.
Includes bibliographical references and index.
ISBN-13: 978-0-226-32140-0 (hardcover : alkaline paper)
ISBN-13: 978-0-226-32142-4 (paperback : alkaline paper)
ISBN-10: 0-226-32140-1 (hardcover : alkaline paper)
ISBN-10: 0-226-32142-8 (paperback : alkaline paper) 1. Digital media—Psychological aspects. 2. Communication and technology. 3. Humanities—Philosophy. 4. Cipher and telegraph codes. 5. Hall, Steven, 1975– Raw shark texts. 6. Danielewski, Mark Z. Only revolutions. I. Title.
P96.T42H39 2012
302.23'1—dc23   2011038467

♾ This paper meets the requirements of ANSI/NISO Z39.48-1992 (Permanence of Paper).

For my family

Contents

Figures

Acknowledgments

If every book is a collaboration—a confluence of thoughts only some of which are original to the author—this one is especially so. In the four or five years this book has been gestating, I have benefited from a great many conversations, books, interviews, comments, and feedback too numerous to mention. Some contributions are so central, however, that I am pleased to acknowledge them explicitly here. First, I thank those scholars who gave generously of their time in agreeing to have an interview with me and for trusting me to handle their comments with fidelity and integrity. These include Eyal Amiran, Jay David Bolter, Tanya Clement, Gregory Crane, Sharon Daniel, Philip J. Ethington, Alice Gambrell, Caren Kaplan, Matthew Kirschenbaum, Timothy Lenoir, Alan Liu, David Lloyd, Tara McPherson, Todd S. Presner, Stephen Ramsey, Rita Raley, and Jeffrey Schnapp. Kenneth Knoespel and Ian Bogost at the School for Literature, Culture and Communication at Georgia Tech provided useful information on their program, its aspirations and priorities. At King's College London, I am especially indebted to Willard McCarty for arranging my visit there, as well as to Harold Short, Stephen Baxter,

Arthur Burns, Hugh Denard, Charlotte Roueche, and Jane Winters for their willingness to allow me to interview them. At the Office of Digital Humanities at the National Endowment for the Humanities, I appreciate the help of Brett Bobley, Jason Rhody, and Jennifer Serventi in talking with me about their programs.

I am especially grateful to Fred Brandes for allowing me to photocopy many of his books and browse through the rest while researching the archival material discussed in chapter 5. He also read through the manuscript and offered several corrections. I warmly remember and deeply appreciate his gracious hospitality, and I thank him for his generosity. I am indebted to the National Cryptologic Museum librarians for their help and encouragement, and I thank them for access to one of the largest intact telegraph code book collections in the United States. The collection was originally assembled when telegraphy was still alive (although long past its heyday), and the National Cryptologic Museum has worked to keep it as a valuable historical archive. I also thank John McVey, who has spent untold hours searching for telegraph code books that have been digitized, primarily in Internet Archive and Google Books, and has created bibliographic records and information for them. He has also compiled a valuable "Resources" page and other assets. I, along with anyone interested in this topic, am indebted to him for his fine and meticulous work. I am grateful as well to Nicholas Gessler for access to his "Things That Think" code book collection and for conversations, information, and explanations that make him my most treasured confidante.

I am indebted to Laura Otis, Marjorie Luesebrink, and Nicholas Gessler for reading drafts and giving valuable feedback. John Johnston, one of the readers of the manuscript for the University of Chicago Press, offered crucial guidance and useful insights that helped enormously in rethinking the book's structure; also invaluable were the second reader's insights and strong endorsement. I owe a large debt of thanks to George Roupe for his meticulous copyediting; this book would be much more error-pone if it had not been for his help. I am also grateful to Alan Thomas of the University of Chicago Press for his continuing guidance and strong support, and to the staff for their good work on this book.

In addition, I thank the National Endowment for the Humanities for support during summer 2009, when I participated in the "Broadening the Humanities" institute led by Tara McPherson and Steve Anderson at the University of Southern California. During the institute I began work on a website to accompany chapter 5 that will have over one hundred code books in digital form with the functionality to do keyword searches, send and receive coded "telegrams" via e-mail, and other resources. George Helfand, then on the staff at USC, helped set up the database for the project, and Zach Blas helped with the website design. I am grateful to them both, and to Steve and Tara for their discussions about the project. I thank Deborah Jakubs, the Rita DiGiallonardo Holloway University Librarian and Vice Provost for Library Affairs, and Winston Atkins at the Duke University Library for assistance in scanning the telegraph code books in the Gessler collection, which will be part of the data repository at the website. A prototype of the website may be found at http://vectorsdev.usc.edu/nehvectors/hayles/site/database.html.

I owe a special debt of gratitude to Allen Beye Riddell, who coauthored the coda to chapter 8 and who did much of the programming on the telegraph code book website. In addition, he also did the research for and authored the website for chapter 8 on Mark Z. Danielewski's *Only Revolutions*. His tireless work on this project has made it much richer and more extensive than it otherwise would have been.

I am grateful to Steven Hall for permission to use images from *The Raw Shark Texts* in chapter 7, as well as for his willingness to read that chapter and offer comments. I am indebted to Mark Z. Danielewski for his friendship and feedback on chapter 8 on *Only Revolutions*, as well as for permission to reprint the images that appear in chapter 8 from *Only Revolutions*.

I am grateful to Joseph Tabbi and Benjamin Robertson at the *Electronic Book Review* for permission to reprint, in revised form, chapter 4, "Tech-TOC: Complex Temporalities in Living and Technical Beings," which appeared in 2011 at http://www.electronicbookreview.com. I am grateful to *Science Fiction Studies* for permission to reprint, in revised form, "Material Entanglements: Steven Hall's *The Raw Shark Texts* as Slipstream Novel," *Science Fiction Studies* 38.1 (March 2011):115–33, which with revisions now appears as chapter 7.

I am grateful to Palgrave Macmillan for permission to use an earlier version of chapter 2, "The Digital Humanities: Engaging the Issues," which appeared in slightly different form in *Understanding Digital Humanities*, edited by David M. Berry (London: Palgrave Macmillan, 2011), reproduced with permission of Palgrave Macmillan. Chapter 3, "How We Read: Close, Hyper, Machine," which appeared in *ADE Bulletin* 150 (2011) and chapter 6, "Narrative and Database: The Limits of Symbiosis," which appeared as "Narrative and Database: Natural Symbionts," *PMLA* 122.5 (October 2007): 1603–8, are reprinted by permission of the Modern Language Association, for which I am grateful. Chapter 8, "Mapping Time, Charting Data: The

Spatial Aesthetic of Mark Z. Danielewski's *Only Revolutions*," appeared in *Essays on Mark Z. Danielewski*, edited by Alison Gibbons and Joe Bray (Manchester: Manchester University Press, 2011), 159–77, and I am grateful to the University of Manchester Press for permission to reprint.

My greatest debt, as always, is to my friend, confidante, and partner, Nicholas Gessler. His comments, feedback, explanations, and help were invaluable. With his extraordinary technical knowledge, his truly amazing collections, and his warm friendship, he has not only made this book much better than it otherwise would be, but he has also made me much better than I otherwise would be.

1

How We Think

Digital Media and Contemporary Technogenesis

How do we think? This book explores the proposition that we think through, with, and alongside media. This, of course, is not a new idea. Marshall McLuhan, Friedrich Kittler, Lev Manovich, Mark Hansen, and a host of others have made similar claims. Building on their work, this book charts the implications of media upheavals within the humanities and qualitative social sciences as traditionally print-based disciplines such as literature, history, philosophy, religion, and art history move into digital media. While the sciences and quantitative social sciences have already made this transition, the humanities and qualitative social sciences are only now facing a paradigm shift in which digital research and publication can no longer be ignored. Starting from mindsets formed by print, nurtured by print, and enabled and constrained by print, humanities scholars are confronting the differences that digital media make in every aspect of humanistic inquiry, including conceptualizing projects, implementing research programs, designing curricula, and educating students. The Age of Print is passing,[1] and the assumptions, presuppositions, and practices associated with it are now becoming visible as media-specific practices rather than the largely invisible status quo.

To evaluate the impact of digital technologies, we may consider in overview an escalating series of effects. At the lower levels are e-mail, departmental websites, web searches, text messaging, creating digital files, saving and disseminating them, and so forth. Nearly everyone in academia, and large numbers outside academia, participate in digital technologies at these levels. Even here, the effects are not negligible. For example, the patterns of errors in writing made with pen and/or typewriter are quite different from those made with word processing. More dramatic is the impact on academic research; whereas scholars used to haunt the library, nowadays they are likely to access the sources they need via web searches. Perhaps most significant at this level is the feeling one has that the world is at one's fingertips. The ability to access and retrieve information on a global scale has a significant impact on how one thinks about one's place in the world. I live in a small town in North Carolina, but thanks to the web, I do not feel in the least isolated. I can access national news, compare it to international coverage, find arcane sources, look up information to fact-check a claim, and a host of other activities that would have taken days in the pre-Internet era instead of minutes, if indeed they could be done at all. Conversely, when my computer goes down or my Internet connection fails, I feel lost, disoriented, unable to work—in fact, I feel as if my hands have been amputated (perhaps recalling Marshall McLuhan's claim that media function as prostheses). Such feelings, which are widespread,[2] constitute nothing less than a change in worldview.

Moreover, research indicates that the small habitual actions associated with web interactions—clicking the mouse, moving a cursor, etc.—may be extraordinarily effective in retraining (or more accurately, repurposing) our neural circuitry, so that the changes are not only psychological but physical as well. Learning to read has been shown to result in significant changes in brain functioning; so has learning to read differently, for example by performing Google searches. Nicholas Carr in *The Shallows: What the Internet Is Doing to Our Brains* (2010) argues that these changes are imperiling our ability to concentrate, leading to superficial thought, diminished capacity to understand complex texts, and a general decline in intellectual capacity. He relates them to feelings of being constantly distracted, so that instead of focusing on a task for a relatively long time, one feels compelled to check e-mail, search the web, break off to play a computer game, and so forth. These issues are discussed in chapter 3, but here I want to draw a somewhat different implication: our interactions with digital media are embodied, and they have bodily effects at the physical level. Similarly, the actions of computers are also embodied, although in a very different manner than with humans. The more one works with digital technologies, the more one comes to appreciate the capacity of networked and programmable machines to carry out sophisticated cognitive tasks, and the more the keyboard comes to seem an extension of one's thoughts rather than an external device on which one types. Embodiment then takes the form of extended cognition, in which human agency and thought are enmeshed within larger networks that extend beyond the desktop computer into the environment. For this reason, models of embodied and extended cognition, such as proposed by Andy Clark (2008) and others, play a central role in my argument.

So far I have been speaking of lower levels of engagement, carried out every day by millions of people. Scholars are among those who frequently enact more sophisticated activities in digital media. At the next level, a scholar begins to use digital technologies as part of the research process. At first this may take the form of displaying results already achieved through other media, for example, posting an essay composed for print on the web. Here the main advantages are worldwide dissemination to a wide variety of audiences, in many cases far beyond what print can reach. The open secret about humanities print publications is their extremely low subscription rates and, beyond this, the shockingly small rate at which articles are cited (and presumably read). David P. Hamilton (1990, 1991) undertook a study of how often journal articles are cited within five years of their publication. Correcting for announcements, reviews, etc., that are not intended for citation (see Pendlebury 1991), his results show that for the sciences, the percentage of articles that have never been cited once in five years is 22.4 percent. For the humanities, it is a whopping 93.1 percent. Even acknowledging the different roles that article publication plays in the sciences (where it is the norm) and the humanities (where the book is the norm) and the different rates at which journal publication takes place in the two fields (a few months in the sciences, from one to three years in the humanities), this figure should give us pause.

The low citation rate suggests that journal publication may serve as a credentialing mechanism for tenure and promotion but that journal publication (with a few significant exceptions) has a negligible audience and a nugatory communicative function. It also raises questions about evaluations of quality. Typically, judgments are made through faculty committees that read a scholar's work and summarize their evaluations for the department. In such deliberations, questions of outreach and audience are rarely entertained in a negative sense (although they are typically considered when work is deemed influential). If influence and audience were considered, one might make a strong argument for taking into account well-written, well-researched blogs that have audiences in the thousands or hundreds of thousands, in contrast to print books and articles that have audiences in the dozens or low hundreds—if that. Indeed, it should make us rethink credentialing in general, as Gary Hall points out in *Digitize This Book! The Politics of New Media or Why We Need Open Access Now* (2008): "The digital model of publishing raises fundamental questions for what scholarly publishing (and teaching) actually is; in doing so it not only poses a threat to the traditional academic hierarchies, but also tells us something about the practices of academic legitimation, authority, judgment, accreditation, and institution in general" (70).

The next step in engagement comes with conceptualizing and implementing research projects in digital media. Here a spectrum of possibilities unfolds: at one end, a one-off project that a scholar undertakes without becoming deeply engaged and, at the other end, scholars who work primarily in digital media. Even at the lower end of the spectrum, assumptions and presuppositions begin to shift in dramatic ways. For example, the scholar who works in digital media is likely to store data in databases rather than express it discursively. As chapter 2 discusses, this change leads to a significant transformation in how a scholar thinks about her material. Refractory elements that must be subordinated in verbal presentation for an argument to make sense and be compelling can now be given weight in their own right. Constructing a database also makes it possible for different scholars (or teams of scholars) to create different front-ends for the same data, thus encouraging collaboration in data collection, storing, and analysis.

At this point the changes accelerate, for now the digital-based scholar begins to shift her perspective more substantially, as issues of design, navigation, graphics, animation, and their integration with concepts come to the fore. While navigation in print is highly constrained, guided by tables of contents, chapter headings, endnotes, indexes, and so on, in web research navigation may occur in a wide variety of ways, each of which has implications for how the audience will encounter and assess the research and thus for what the research is taken to mean. Hypertext links, hierarchies of screen displays, home page tabs, and so forth all contribute to the overall effect. Graphics, animation, design, video, and sound acquire argumentative force and become part of the research's quest for meaning. As a scholar confronts these issues, sooner or later she will likely encounter the limits of her own knowledge and skills and recognize the need—indeed, the necessity—for collaboration. Since the best collaborations are those in which all the partners are in from the beginning and participate in the project's conceptualization as well as implementation, this in turn implies a very different model of work than the typical procedures of a print-based scholar, who may cooperate with others in a variety of ways, from citing other scholars to asking acquaintances to read manuscripts, but who typically composes alone rather than in a team environment.

Working collaboratively, the digitally based scholar is apt to enlist students in the project, and this leads quickly to conceptualizing courses in which web projects constitute an integral part of the work. Now the changes radiate out from an individual research project into curricular transformation and, not coincidentally, into different physical arrangements of instruction and research space. The classroom is no longer sufficient for the needs of web pedagogy; needed are flexible laboratory spaces in which teams can work collaboratively, as well as studio spaces with high-end technologies for production and implementation. At this point, it is difficult to say where the transformations end, for now almost every aspect of work in the humanities can be envisioned differently, including research and publication, teaching and mentoring, credentialing and peer evaluation, and last but not least, relations of the academy to the larger society.

Such wide-ranging shifts in perspective often are most dramatically evident in scholars who have administrative responsibility, represented in this study (discussed in chapter 2) by Kenneth Knoespel at Georgia Tech; Tara McPherson at the University of Southern California; Alan Liu at the University of California, Santa Barbara; Harold Short at King's College London; and Jeffrey Schnapp (who was at Stanford University when I interviewed him but has since moved to Harvard University). As administrators, they must necessarily think programmatically about where their administrative units are going, how present trends point to future possibilities, how outcomes will be judged, and how their units relate to the university and the society in general. They clearly understand that digital technologies, in broad view, imply transformation not only of the humanities but of the entire educational system. They are also keenly aware of difficulties to be negotiated within the humanities as traditionally print-based disciplines fracture into diverse contingents, with some scholars still firmly within the regime of print while others are racing into the digital domain.

The changes charted here have been represented as a series of levels with gradual increases between them. However, if the lowest level is compared directly with the highest, the differences are stark, pointing to the possibility of a widening rift between print- and digital-based scholars. This situation poses a host of theoretical, organizational, and pedagogical challenges. As the Digital Humanities mature, scholars working within digital media are developing vocabularies, rhetorics, and knowledge bases necessary for the advancement of the field. To a certain extent, knowledge construction is cumulative, and the citations, allusions, and specialized discourses of the Digital Humanities presume audiences capable of contextualizing and understanding the stakes of an argument; the implications of a project; the innovations, resistances, and disruptions that research strategies pose to work that has gone before. At the same time, however, traditional (i.e., print-based) scholars are struggling to grasp the implications of this work and often failing to do so.

The failures are apt to take two distinct but related forms. First, print-based scholars are inclined to think that the media upheavals caused by the advent of digital technologies are no big deal. In this view, digital text is read as if it were print, an assumption encouraged by the fact that both books and computer screens are held at about the same distance from the eyes. Moreover, print-based scholars increasingly compose, edit, and disseminate files in digital form without worrying too much about how digital text differs from print, so they tend not to see the ways in which digital text, although superficially similar to print, differs profoundly in its internal structures, as well as in the different functionalities, protocols, and communicative possibilities of networked and programmable machines. The second kind of failure manifests as resistance to, or outright rejection of, work in digital media. Many factors are implicated in these responses, ranging from anxieties that (print) skill sets laboriously acquired over years of effort may become obsolete, to judgments formed by print aesthetics that undervalue and underrate digital work, leading to a kind of tunnel vision that focuses on text to the exclusion of everything else such as graphics, animation, navigation, etc.

Faced with these resistances and misunderstandings, humanities scholars working in digital media increasingly feel that they are confronted with an unsavory dilemma: either they keep trying to explain to their print-based colleagues the nature and significance of their work, fighting rearguard actions over and over at the expense of developing their own practices, or else they give up on this venture, cease trying to communicate meaningfully, and go their own way. The resulting rift between print-based and digital scholarship would have significant implications for both sides. Print-based scholars would become increasingly marginalized, unable to communicate not only with Digital Humanities colleagues but also with researchers in the social sciences and sciences, who routinely use digital media and have developed a wide range of skills to work in them. Digital humanities would become cut off from the rich resources of print traditions, leaving behind millennia of thought, expression, and practice that no longer seem relevant to its concerns.

Surely there must be a better way. Needed are approaches that can locate digital work within print traditions, and print traditions within digital media, without obscuring or failing to account for the differences between them. One such approach is advocated here: it goes by the name of Comparative Media Studies.[3] As a concept, Comparative Media Studies has long inhabited the humanities, including comparisons of manuscript and print cultures, oral versus literate cultures, papyri versus vellum, immobile type versus moveable type, letterpress versus offset printing, etc. These fields have tended to exist at the margins of literary culture, of interest to specialists but (with significant exceptions) rarely sweeping the humanities as a whole. Moreover, they have occupied separate niches without overall theoretical and conceptual frameworks within which Comparative Media Studies might evolve.

With the momentous shift from print to digital media within the humanities, Comparative Media Studies provides a rubric within which the interests of print-based and digital humanities scholars can come together to explore synergies between print and digital media, at the same time bringing into view other versions of Comparative Media Studies, such as the transition from manuscript to print culture, that have until now been relegated to specialized subfields. Building on important work in textual and bibliographic studies, it emphasizes the importance of materiality in media. Broadening the purview beyond print, it provides a unifying framework within which curricula may be designed systematically to initiate students into media regimes, highlighting the different kinds of reading practices, literacies, and communities prominent in various media epochs.

Examples of Comparative Media Studies include research that combines print and digital literary productions, such as Matthew Kirschenbaum's (2007) concepts of formal and forensic materiality, Loss Glazier's (2008) work on experimental poetics, John Cayley (2004, 2002) on letters and bits, and Stephanie Strickland (Strickland 2002; Strickland and Lawson 2002) on works that have both print and digital manifestations. Other examples are theoretical approaches that combine continental philosophy with New Media content, such as Mark Hansen's *New Philosophy for New Media* (2006b).

Still others are provided by the MIT series on platform studies, codirected by Nick Montfort and Ian Bogost (Montfort and Bogost 2009), which aims to locate specific effects in the affordances and constraints of media platforms such as the Atari 5600 video game system, in which the techniques of close reading are applied to code and video display rather than text. Also in this grouping are critical code studies, initiated by Wendy Hui Kyong Chun (2008, 2011) and Mark Marino (2006) among others, that bring ideology critique to the rhetoric, form, and procedures of software. In this vein as well is Ian Bogost's work (2007) on procedural rhetorics, combining traditional rhetorical vocabularies and approaches with software functionalities. Lev Manovich's recent (2007) initiative, undertaken with Jeremy Douglas, on "cultural analytics" uses statistical analysis and database structures to analyze large data sets of visual print materials, such as *Time* covers from 1923 to 1989, and one million pages of manga graphic novels (discussed in chapter 3). Diverse as these projects are, they share an assumption that techniques, knowledges, and theories developed within print traditions can synergistically combine with digital productions to produce and catalyze new kinds of knowledge.

On a pedagogical level, Comparative Media Studies implies course designs that strive to break the transparency of print and denaturalize it by comparing it with other media forms. Alan Liu (2008c) at the University of California, Santa Barbara, has devised a series of courses that he calls "Literature+" (discussed in chapter 3), which combines close reading of print texts with comparisons to other media forms. Another example is a seminar comparing the transition from manuscript to print with that of print to digital, offered at Yale University by Jessica Brantley, a medievalist, and Jessica Pressman, a specialist in contemporary literature. Other approaches might stress multiple literacies that include print but also emphasize writing for the web, designing computer games, creating simulations of social situations, and a variety of other media modalities. My colleagues at Duke University, including Cathy Davidson, Nicholas Gessler, Mark Hansen, Timothy Lenoir, and Victoria Szabo, are creating courses and research projects that follow such interdisciplinary lines of inquiry. Extrapolating from these kinds of experiments, Comparative Media Studies can provide a framework for courses in which students would acquire a wide repertoire of strategies to address complex problems. Faced with a particular kind of problem, they would not be confined to only one mode of address but could think creatively about the resources, approaches, and strategies the problem requires and choose the more promising one, or an appropriate combination of two or more, for a given context.

Such a curriculum is worlds away from the offerings of a traditional English department, which typically focuses on periodizations (e.g., eighteenth century prose), nationalities (British, American, Anglophone, etc.), and genres (fiction, prose, drama). The difficulties with this kind of approach are not only that it is outmoded and fails to account for what much of contemporary scholarship is about (postcolonial studies, globalization studies, race and gender studies, etc.). It also focuses on content rather than problems, assuming that students will somehow make the leap from classroom exercises to real-world complexities by themselves. To be sure, not every intellectual exercise may be framed as a problem. The humanities have specialized in education that aims at enriching a student's sense of the specificity and complexity of our intellectual heritage, including major philosophical texts, complex literary works, and the intricate structures of theoretical investigations into language, society, and the human psyche. Nevertheless, there must also be a place for problem-based inquiry within the humanities as well as the sciences and social sciences. Comparative Media Studies is well suited to this role and can approach it through the framework of multiple literacies.

The implications of moving from content orientation to problem orientation are profound. Project-based research, typical of work in the Digital Humanities, joins theory and practice through the productive work of making. Moreover, the projects themselves evolve within collaborative environments in which research and teaching blend with one another in the context of teams with many different kinds of skills, typically in spaces fluidly configured as integrated classroom, laboratory, and studio spaces. The challenges of production complicate and extend the traditional challenges of reading and writing well, adding other dimensions of software utilization, analytical and statistical tools, database designs, and other modalities intrinsic to work in digital media. Without abandoning print literacy, Comparative Media Studies enriches it through judicious comparison with other media, so that print is no longer the default mode into which one falls without much thought about alternatives but rather an informed choice made with full awareness of its possibilities and limitations. Conceptualized in this way, Comparative Media Studies courses would have wide appeal not only within the humanities but in the social sciences and some of the hard sciences as well. Such courses would provide essential preparation for students entering the information-intensive and media-rich environments in which their careers will be forged and their lives lived.

Adopting this perspective requires rethinking priorities and assumptions on so many levels that it is more like peeling an onion than arriving at a decision. One thinks one understands the implications, but then further layers reveal themselves and present new challenges to the scholar who has grown up with print, taught with print, and conducted research exclusively in print media. A principal aim of this book is to excavate these layers, showing through specific case studies what Comparative Media Studies involves. One way into the complexities is to track the evolution of the Digital Humanities, the site within the humanities where the changes are most apparent and, arguably, most disruptive to the status quo. As chapter 2 shows, the Digital Humanities are not a monolithic field but rather a collection of dynamic evolving practices, with internal disputes, an emerging set of theoretical concerns interwoven with diverse practices, and contextual solutions to specific institutional configurations.

Another way is through the concept of technogenesis, the idea that humans and technics have coevolved together. The proposition that humans coevolved with the development and transport of tools is not considered especially controversial among paleoanthropologists. For example, the view that bipedalism coevolved with tool manufacture and transport is widely accepted. Walking on two legs freed the hands, and the resulting facility with tools bestowed such strong adaptive advantage that the development of bipedalism was further accelerated, in a recursive upward spiral that Andy Clark (2008) calls "continuous reciprocal causation." To adapt this idea to the contemporary moment, two modifications are necessary. The first was proposed in the late nineteenth century by James Mark Baldwin (1896), now referred to as the Baldwin effect. He suggested that when a genetic mutation occurs, its spread through a population is accelerated when the species reengineers its environment in ways that make the mutation more adaptive. Updating Baldwin, recent work in evolutionary biology has acknowledged the importance of epigenetic changes—changes initiated and transmitted through the environment rather than through the genetic code. This allows for a second modification, the idea that epigenetic changes in human biology can be accelerated by changes in the environment that make them even more adaptive, which leads to further epigenetic changes. Because the dynamic involves causation that operates through epigenetic changes, which occur much faster than genetic mutations, evolution can now happen much faster, especially in environments that are rapidly transforming with multiple factors pushing in similar directions. Lending credence to this hypothesis is recent work in neurophysiology, neurology, and cognitive science, which has shown that the brain, central nervous system, and peripheral nervous system are endowed with a high degree of neural plasticity. While greatest in infants, children, and young people, neural plasticity continues to some extent into adulthood and even into old age.

As digital media, including networked and programmable desktop stations, mobile devices, and other computational media embedded in the environment, become more pervasive, they push us in the direction of faster communication, more intense and varied information streams, more integration of humans and intelligent machines, and more interactions of language with code. These environmental changes have significant neurological consequences, many of which are now becoming evident in young people and to a lesser degree in almost everyone who interacts with digital media on a regular basis.

The epigenetic changes associated with digital technologies are explored in chapter 3 through the interrelated topics of reading and attention. Learning to read complex texts (i.e., "close reading") has long been seen as the special province of the humanities, and humanities scholars pride themselves on knowing how to do it well and how to teach students to do it. With the advent of digital media, other modes of reading are claiming an increasing share of what counts as "literacy," including hyper reading and analysis through machine algorithms ("machine reading"). Hyper reading, often associated with reading on the web, has also been shown to bring about cognitive and morphological changes in the brain. Young people are at the leading edge of these changes, but pedagogical strategies have not to date generally been fashioned to take advantage of these changes. Students read and write print texts in the classroom and consume and create digital texts of their own on screens (with computers, iPhones, tablets, etc.), but there is little transfer from leisure activities to classroom instruction or vice versa. A Comparative Media Studies perspective can result in courses and curricula that recognize all three reading modalities—close, hyper-, and machine—and prepare students to understand the limitations and affordances of each.

Fred Brooks, a computer scientist at the University of North Carolina and author of the best-selling *The Mythical Man-Month* (alluding to the flawed assumption that more manpower inevitably means faster progress), offers good advice relevant to crafting a Comparative Media Studies approach in *The Design of Design: Essays from a Computer Scientist* (2010a). In an interview in *Wired*, he comments that "the critical thing about the design process is to identify your scarcest resource. Despite what you may think, that very often is not money. For example, in a NASA moon shot, money is abundant but lightness is scarce; every ounce of weight requires tons of material below. On the design of a beach vacation home, the limitation may be your oceanfront footage. You have to make sure your whole team understands what scarce resource you're optimizing" (2010b:92). The answer to the "scarce resource" question for societies in developed countries seems clear: the sheer onslaught of information has created a situation in which the limiting factor is human attention. There is too much to attend to and too little time to do it. (The situation is of course quite different in developing countries, where money may indeed function as the scarce resource.)

Hyper reading, which includes skimming, scanning, fragmenting, and juxtaposing texts, is a strategic response to an information-intensive environment, aiming to conserve attention by quickly identifying relevant information, so that only relatively few portions of a given text are actually read. Hyper reading correlates, I suggest, with hyper attention, a cognitive mode that has a low threshold for boredom, alternates flexibly between different information streams, and prefers a high level of stimulation. Close reading, by contrast, correlates with deep attention, the cognitive mode traditionally associated with the humanities that prefers a single information stream, focuses on a single cultural object for a relatively long time, and has a high tolerance for boredom. These correlations suggest the need for pedagogical strategies that recognize the strengths and limitations of each cognitive mode; by implication, they underscore the necessity for building bridges between them. Chapter 3, where these matters are discussed, begins weaving the thread of attention/distraction that runs throughout the book. If we think about humanities research and teaching as problems in design (i.e., moving from content orientation to problem orientation), then Brooks's advice suggests that for collaborative teams working together to craft projects and curricula in digital media, it is crucial for team partners to recognize the importance of human attention as a limiting/enabling factor, both as a design strategy and as a conceptual framework for theoretical work. In an academic context, of course, the issue is not as simple as optimization, for pedagogical goals and research projects may aim at disruption and subversion rather than replication. This caveat notwithstanding, attention as a focus for inquiry opens onto a complex and urgent set of issues, including the relation of human to machine cognition and the cycles of epigenetic changes catalyzed by our increasing exposure to and engagement with digital media.

To flesh out the concept of technogenesis and to explore how a technology platform can initiate wide-ranging changes in society, chapter 5 undertakes a case study of the first globally pervasive binary signaling system, the telegraph. The focus is on telegraph code books, print productions that offered "economy, secrecy and simplicity" by matching natural-language phrases with corresponding code words. Affecting the wider society through the changes that telegraphy catalyzed, telegraph code books demonstrate that changed relations of language and code, bodily practices and technocratic regimes, and messages and cultural imaginaries created technogenetic feedback loops that, over the course of a century, contributed significantly to reengineering the conditions of everyday life. In this sense, telegraphy anticipated the epigenetic changes associated with digital technologies, especially fast communication and the virtualization of commodities.

When humanities scholars turn to digital media, they confront technologies that operate on vastly different time scales, and in significantly different cognitive modes, than human understanding. Grasping the complex ways in which the time scales of human cognition interact with those of intelligent machines requires a theoretical framework in which objects are seen not as static entities that, once created, remain the same throughout time but rather are understood as constantly changing assemblages in which inequalities and inefficiencies in their operations drive them toward breakdown, disruption, innovation, and change. Objects in this view are more like technical individuals enmeshed in networks of social, economic, and technological relations, some of which are human, some nonhuman. Among those who have theorized technical objects in this way are Gilbert Simondon, Adrian Mackenzie, Bruno Latour, and Matthew Fuller. Building on their work, I hypothesize in chapter 4 about the multilevel, multiagent interactions occurring across the radically different time scales in which human and machine cognitions intermesh: on the human side, the very short time scales of synaptic connections to the relatively long time scales required for narrative comprehension; on the machine side, the very fast processing at the level of logic gates and bit reading to the relatively long load times of complex programs. Obviously, the meshing of these two different kinds of complex temporalities does not happen all at one time (or all at one place) but rather evolves as a complex syncopation between conscious and unconscious perceptions for humans, and the integration of surface displays and algorithmic procedures for machines. The interactions are dynamic and continuous, with feedback and feedforward loops connecting different levels with each other and cross-connecting machine processes with human responses.

On the level of conscious thought, attention comes into play as a focusing action that codetermines what we call materiality. That is, attention selects from the vast (essentially infinite) repertoire of physical attributes some characteristics for notice, and they in turn constitute an object's materiality. Materiality, like the object itself, is not a pre-given entity but rather a dynamic process that changes as the focus of attention shifts. Perceptions exist unconsciously as well as consciously, and research emerging from contemporary neuroscience, psychology, and other fields about the "new unconscious" (or "adaptive unconscious") plays a critical role in understanding this phenomenon. In these views, the unconscious does not exist primarily as repressed or suppressed material but rather as a perceptive capacity that catches the abundant overflow too varied, rich, and deep to make it through the bottleneck of attention. Attention, as the limiting scarce resource, directs conscious notice, but it is far from the whole of cognitive activity and in fact constitutes a rather small percentage of cognition as a whole. The realization that neural plasticity happens at many levels, including unconscious perceptions, makes technogenesis a potent site for constructive interventions in the humanities as they increasingly turn to digital technologies. Comparative Media Studies, with its foregrounding of media technologies in comparative contexts, provides theoretical, conceptual, and practical frameworks for critically assessing technogenetic changes and devising strategies to help guide them in socially constructive ways.

If time is deeply involved with the productions of digital media, so too is space. GIS (geographic information system) mapping, GPS (global positioning system) technologies, and their connections with networked and programmable machines have created a culture of spatial exploration in digital media. At least as far back as Henri Lefebvre's *The Production of Space* ([1974] 1992), contemporary geographers have thought about space not in static Cartesian terms (which Lefebvre calls represented or conceived space) but as produced through networks of social interactions. As Lefebvre proclaims, (social) practices produce (social) spaces. Among contemporary geographers, Doreen Massey (1994a, 1994b, 2005) stands out for the depth of her research and intelligent advocacy of an approach to social spaces based on interrelationality, open-ended temporality, and a refusal of space represented as a Cartesian grid. For spatial history projects, however, georeferencing relational databases to the "absolute space" of inches, miles, and kilometers has proven unavoidable and indeed desirable, since it allows interoperability with the data sets and databases of other researchers. The tensions between Massey's dream (as it is called in chapter 6) and the spatial history projects exemplified by the Stanford Spatial History Project show the limitations as well as the theoretical force of Massey's approach.

The inclusion of databases in spatial history projects has opened the door to new strategies that, rather than using narrative as their primary mode of explication, allow flexible interactions between different layers and overlays. As a result, explanations move from charting linear chains of causes and effects to more complex interactions among and between networks located in space and time. Moreover, historical projects have also moved from relational databases, in which data elements are coordinated through shared keys (i.e., common data elements), to object-oriented databases, in which classes possess inheritable traits and aggregative potentials. As Michael Goodchild (2008) explains, the older relational model implies a metaphor of GIS as a container of maps. One constructs a map by merging different data elements into a common layer. While this strategy works well for certain kinds of explanations, it has the disadvantage of storing data in multiple databases and creating spatial displays that have difficulty showing change through time. Newer object-oriented databases, by contrast, imply a metaphor of objects in the world that can spawn progeny with inherited traits, merge with other objects, and aggregate into groups. This makes it possible to chart their movements through time in ways that make time an intrinsic property rather than something added on at the end by marking layers with time indicators.

Whereas historical and historically inflected projects are finding new ways to construct and display social space, experimental literature plays with the construction of imaginary spaces. Chapter 7 explores Steven Hall's distributed literary system that has as its main component the print novel *The Raw Shark Texts: A Novel* ([2007] 2008a). In depicting a posthuman subjectivity that has transformed into a huge online database capable of evacuating individual subjectivities and turning them into "node bodies," the text performs a critique of postindustrial knowledge work as analyzed by Alan Liu (2008b). In the print text, the distance between signifier and signified collapses, so that letters form not only words but also objects and living beings. In the "unspace" of abandoned tunnels, warehouses, and cellars, the story evolves of amnesiac Eric Sanderson's search for his past memories while he is pursued by a "conceptual shark," the Lodovician, which hunts him through the trails of thoughts, perceptions, and memories that he emits. While social space is constructed through social practices, "unspace" is constructed through words that at once signify and function as material objects. The materiality of language is here given a literal interpretation, and the resulting conflation of imaginary with physical space creates an alternative universe mapped as well as denoted by language. Supremely conscious of itself as a print production, this book explores the linguistic pleasures and dangerous seductions of immersive fictions, while at the same time exploring the possibilities for extending its narrative into transmedial productions at Internet sites, translations into other languages, and physical locations.

With the advent of digital databases and the movement of traditionally narrative fields such as qualitative history into new kinds of explanations and new modes of data displays, narrative literature has fashioned its own responses to information-intensive environments. As Lev Manovich has noted, narrative and database have complementary strengths and limitations (2002:190–212). Narrative excels in constructing causal models, exploiting complex temporalities, and creating models of how (other) minds work. Databases, by contrast, specialize in organizing data into types and enabling the flexible concatenation of data elements. In an era when databases are perhaps the dominant cultural form, it is no surprise that writers are on the one hand resisting databases, as *The Raw Shark Texts* (2008e) does, and on the other hand experimenting with ways to combine narrative and database into new kinds of literature, as does Mark Z. Danielewski's *Only Revolutions* (2007b). Part epic poem, part chronological database of historical events, *Only Revolutions* pushes the envelope of literary forms that may still be called "a novel."

One of the ways in which *Only Revolutions* works, discussed in chapter 8, is through the application of an extensive set of constraints, mirroring in this respect the structured forms of relational databases and database queries. Whereas relational databases allow multiple ways to concatenate data elements, the spatial aesthetic of *Only Revolutions* creates multiple ways to read every page spread by dividing the page into clearly delineated sections that can be cross-correlated. Moreover, an invisible constraint governs the discourse of the entire text—Danielewski's previous novel *House of Leaves* (2000), which functions as a mirror opposite to *Only Revolutions.* Whatever was emphasized in *House of Leaves* is forbidden to appear in *Only Revolutions*, so that what cannot be spoken or written becomes a powerful force in determining what *is* written or spoken. In this sense, *Only Revolutions* posits an Other to itself that suggests two responses to the information explosion: a novel that attempts to incorporate all different kinds of discourses, sign systems, and information into itself, engorging itself in a frenzy of graphomania (i.e., *House of Leaves*) and a novel that operates through severe constraints, as if keeping the information deluge at bay through carefully constructed dikes and levees (i.e., *Only Revolutions*). In the first case, attention is taxed to the limit through writing strategies that fill and overfill the pages; in the second case, attention is spread among different textual modalities, each interacting with and constraining what is possible in the others.

In conclusion, I offer a few reflections on my book title and on the book as itself a technogenetic intervention. *How We Think* encompasses a diverse sense of "we," focusing in particular on the differences and overlaps between the perspectives of print-based and digital-based scholars in the humanities and qualitative social sciences. "Think"—a loaded word if ever there was one—implies in this context both conscious and unconscious perceptions, as well human and machine cognition. Like humans, objects also have their embodiments, and their embodiments matter, no less than for humans. When objects acquire sensors and actuators, it is no exaggeration to say they have an *umwelt*, in the sense that they perceive the world, draw conclusions based on their perceptions, and act on those perceptions.[4] All this takes place, of course, without consciousness, so their modes of being in the world raise deep questions about the role of consciousness in embodied and extended cognition. The position taken throughout this book is that all cognition is embodied, which is to say that for humans, it exists throughout the body, not only in the neocortex. Moreover, it extends beyond the body's boundaries in ways that challenge our ability to say where or even if cognitive networks end.

Making the case for technogenesis as a site for constructive interventions, this book performs the three reading strategies discussed in chapter 3 of close, hyper-, and machine reading. The literary texts discussed here provide the occasion for close reading. Since these texts are deeply influenced by digital technologies, they are embedded in information-intensive contexts that require and demand hyper reading, which in conjunction with close reading provided the wide range of references used throughout the book. Finally, the coda to chapter 8, written in collaboration with Allen Riddell, presents results from our machine reading of *Only Revolutions*. Combining close, hyper-, and machine reading with a focus on technogenesis, the book is meant as a proof of concept of the potential of Comparative Media Studies not only in its arguments but also in the methodologies it instantiates and the interpretive strategies it employs.

Momentous transformations associated with digital technologies have been recognized and documented by a plethora of studies discussing economic, social, political, and psychological changes. However, people are the ones driving these changes through myriad decisions about how to use the technologies. This lesson was clear at the very beginning of the Internet, when users grasped its potential for communication and especially the usefulness of web browsers for expression and display. Every major development since then has been successful not (or not only) because of intrinsic technological capability but because users found ways to employ them to pursue their own interests and goals. Hacktivism, the open source movement, user listservs, music and video file sharing, social networking, political games, and other practices in digital media are user-driven and often user-defined; they are potent forces in transforming digital technologies so that they become more responsive to social and cultural inequities, more sensitive to webs of interconnections between people and between people and objects, more resistant to predatory capitalistic practices. In this view, digital media and contemporary technogenesis constitute a complex adaptive system, with the technologies constantly changing as well as bringing about change in those whose lives are enmeshed with them.

We are now in a period when the interests of individuals are in dynamic interplay with the vested interests of large corporations, sometimes working together to create win-win situations, other times in sharp conflict over whose interests will prevail. Contemporary technogenesis encompasses both possibilities, as well as the spectrum of other outcomes in between; as a phrase, it does not specify the direction or human value of the changes, whether for good or ill. This book takes that ambiguity as its central focus, as it attempts to intervene in locally specific ways in the media upheavals currently in progress by showing how digital media can be used fruitfully to redirect and reinvigorate humanistic inquiry. People—not the technologies in themselves—will decide through action and inaction whether an intervention such as this will be successful. In this sense, my title is as much an open-ended question as an assertion or claim.

FIRST INTERLUDE

Practices and Processes in Digital Media

The idea of practice-based research, long integrated into the sciences, is relatively new to the humanities. The work of making—producing something that requires long hours, intense thought, and considerable technical skill—has significant implications that go beyond the crafting of words. Involved are embodied interactions with digital technologies, frequent testing of code and other functionalities that results in reworking and correcting, and dynamic, ongoing discussions with collaborators to get it right. As Andy Pickering has cogently argued in *The Mangle of Practice: Time, Agency, and Science* (1995), practice as embodied skill is intimately involved with conceptualization. Conceptualization suggests new techniques to try, and practices refine and test concepts, sometimes resulting in significant changes in how concepts are formulated.

Coming to the scene with a background in scientific programming and a long-standing interest in machine cognition, I wanted to see how engagements with digital media are changing the ways in which humanities scholars think. The obvious and visible signs of a shift include the changing nature of research, the inclusion of programming code as a necessary linguistic practice, and the increasing number of large web projects in nearly every humanities discipline. This much I knew, but I was after something deeper and more elusive: how engagements with digital technologies are affecting the presuppositions and the assumptions of humanities scholars, including their visions of themselves as professional practitioners, their relations to the field, and their hopes and fears for the future. I hoped that by talking with researchers directly in interviews and site visits, I would get more nuanced, candid, and intuitive responses than I might access through publications. The desire to get below the surface, so to speak, will be given theoretical weight in subsequent chapters; in chapter 2, it comes in the form of wide-bandwidth communication with researchers, with many of whom I had long-standing relationships and shared a degree of trust.

Chapter 2 shows a widening circle of effects that, in their most extensive form, imply nothing less than a complete rethinking of almost every aspect of the humanities. This raises the urgent question of how the Traditional Humanities and the Digital Humanities will relate to one another. While it is possible each will choose to go its own way separate from the other, I argue that this would be a tragic mistake. Rather, the Comparative Media Studies approach discussed in chapter 1 provides a way to integrate concerns central to print-based humanities with the Digital Humanities, without neglecting the legitimate concerns of each and while building a framework in which each can catalyze new insights, research questions, and theoretical agendas for the other.

Processes, closely related to but distinct from practices, imply evolution through time, including temporary concrescences that can take forms both internal and external to the human body and can include technical objects as well as living beings. Whole-body interactions discussed in chapter 3 include neurological changes that have the effect of rewiring the brain as a result of interacting with digital media, especially reading (and writing) on the web. My methodology here includes extensive reading in neurological research, a practice that many humanists would prefer not to follow, for good reasons. They argue that relying on such research automatically places humanists at a disadvantage, for humanistic training does not equip us to evaluate such work responsibly. Moreover, they feel that turning to the sciences has the effect of placing the humanities in a subsidiary or inferior position, as if we must rely on scientists to tell us what we think (or more to the point here, how we think). In addition, there is the unsettling tendency of the scientific "truths" of one generation to be overturned or reversed in the next, especially in fields as rapidly developing as brain science and the science of consciousness. Granting the cogency of these objections, I nevertheless believe that in this instance, humanists have as much or more experience as anyone in how people read and in evaluating the consequences (neurological as well as psychological, political, theoretical) of their reading. Our own experiences can provide useful grounding from which to evaluate the scientific research. Turning specifically to how reading is taught in English and literature departments, I argue that the traditional emphasis on close reading should be enlarged with attention to hyper reading and machine reading. Young people practice hyper reading extensively when they read on the web, but they frequently do not use it in rigorous and disciplined ways. Incorporating such instruction in the literature classroom and making explicit connections with close reading, as well as enlarging the repertoire by including machine reading, offers a potent solution to the national crisis in reading that many surveys detect. The theme of attention introduced in this chapter will continue as a connecting thread throughout the book.

2

The Digital Humanities

Engaging the Issues

Arguably more print-based than the sciences and social sciences, the humanities are also experiencing the effects of digital technologies. At the epicenter of change are the Digital Humanities. The Digital Humanities have been around since at least the 1940s,[1] but it was not until the Internet and World Wide Web that they came into their own as emerging fields with their own degree programs, research centers, scholarly journals and books, and a growing body of expert practitioners. Nevertheless, many humanities scholars remain only vaguely aware of the Digital Humanities and lack a clear sense of the challenges they pose to traditional modes of inquiry. This chapter outlines the field, analyzes the implications of its practices, and discusses its potential for transforming research, teaching, and publication. As a subversive force, the Digital Humanities should not be considered a panacea for whatever ails the humanities, for they brings their own challenges and limitations. The point, to my mind, is not that it is better (or worse) but rather that it is *different*, and the differences can leverage traditional assumptions so they become visible and hence available for rethinking and reconceptualizing.

To explore these issues, I conducted a series of phone and in-person interviews with twenty US scholars at different stages of their careers and varying intensities of involvement with digital technologies. I also made site visits to the Centre for Computing in the Humanities (CCH) at King's College London and to the School for Literature, Culture and Communication at Georgia Tech. In addition, I conducted an interview with the program officers at the Office of Digital Humanities at the National Endowment for the Humanities (NEH). The insights that my interlocutors expressed in these conversations were remarkable. Through narrated experiences, sketched contexts, subtle nuances, and implicit conclusions, the interviews reveal the ways in which the Digital Humanities are transforming assumptions. The themes that emerged can be grouped under the following rubrics: scale, critical/productive theory, collaboration, databases, multimodal scholarship, code, and future trajectories. As we will see, each of these areas has its own tensions, conflicts, and intellectual issues. I do not find these contestations unsettling; on the contrary, I think they indicate the vitality of the Digital Humanities and their potential for catalyzing significant change.

Defining the Field

Nowhere is this contentious vitality more evident than in the field's definition. The rapid pace of technological change correlates in the Digital Humanities with different emphases, opening a window onto the field's history, the controversies that have shaped it, and the tensions that continue to resonate through it. Stephen Ramsay (2008b) recalls, "I was present when this term was born . . . 'digital humanities' was explicitly created—by Johanna Drucker, John Unsworth, Jerome McGann, and a few others who were at IATH [Institute for Advanced Technology in the Humanities at the University of Virginia] in the late nineties—to replace the term 'humanities computing.' The latter was felt to be too closely associated with computing support services, and for a community that was still young, it was important to get it right." Alan Liu (2008a) also recalls using the term around 1999–2000. Although some practitioners continue to prefer "humanities computing,"[2] for Ramsay and his colleagues, "Digital Humanities" was meant to signal that the field had emerged from the low-prestige status of a support service into a genuinely intellectual endeavor with its own professional practices, rigorous standards, and exciting theoretical explorations. On this last point, Matthew Kirschenbaum (2009) recalls the convergence of the Digital Humanities with newly revitalized bibliographic studies, as Jerome McGann and others were challenging traditional wisdom and advocating for a contextualized cultural studies approach: "The combat in the editorial community . . . provided first-wave Digital Humanities with a theoretical intensity and practical focus that would have been unattainable had we simply been looking at digitization and database projects broadly construed. . . . The silver bullet of first-wave Digital Humanities, it seems to me, was the conjoining of a massive theoretical shift in textual studies with the practical means to implement and experiment with it." Kirschenbaum (2010) also traces the term to the search for an appropriate title for the Blackwell *Companion to Digital Humanities* by John Unsworth in conversation with Andrew McNellie, a Blackwell editor. In addition, he cites Brett Bobley at the NEH Office of Digital Humanities deciding it was superior to "humanities computing" because it "implied a form of humanism" (qtd. in Kirschenbaum 2010).

A decade later, the term is morphing again as some scholars advocate a turn from a primary focus on text encoding, analysis, and searching to multimedia practices that explore the fusion of text-based humanities with film, sound, animation, graphics, and other multimodal practices across real, mixed, and virtual reality platforms. The trajectory can be traced by comparing John Unsworth's "What Is Humanities Computing and What is Not?" (2002) with Jeffrey Schnapp and Todd Presner's "The Digital Humanities Manifesto 2.0" (Schnapp and Presner 2009). At the top of Unsworth's value hierarchy are sites featuring powerful search algorithms that offer users the opportunity to reconfigure them to suit their needs. Sites billing themselves as Digital Humanities but lacking the strong computational infrastructure are, in Unsworth's phrase, "charlatans."

By contrast, the "Manifesto" consigns values such as Unsworth's to the first wave, asserting that it has been succeeded by a second wave emphasizing user experience rather than computational design:

> The digital first wave replicated the world of scholarly communications that print gradually codified over the course of five centuries: a world where textuality was primary and visuality and sound were secondary. . . . Now it must shape a future in which the medium-specific features of digital technologies become its core and in which print is absorbed into new hybrid modes of communication.
>
> The first wave of digital humanities work was quantitative, mobilizing the search and retrieval powers of the database, automating corpus linguistics,

> stacking hypercards into critical arrays. The second wave is **qualitative, interpretive, experiential, emotive, generative** in character [boldface in original]. It harnesses digital toolkits in the service of the Humanities' core methodological strengths: attention to complexity, medium specificity, historical context, analytical depth, critique and interpretation.

Note that the core mission is here defined so that it no longer springs primarily from quantitative analyses of texts but rather from practices and qualities that can inhere in any medium. In this view the Digital Humanities, although maintaining ties with text-based study, have moved much closer to time-based art forms such as film and music, visual traditions such as graphics and design, spatial practices such as architecture and geography, and curatorial practices associated with museums, galleries, and the like.[3] Understandably, pioneers of the so-called first wave do not unequivocally accept this characterization, sometimes voicing the view that "second wave" advocates are Johnny-come-latelies who fail to understand what the Digital Humanities really are.

In a quantitative/qualitative analysis of this tension, Patrick Svensson (2009) shows that the tradition of textual analyses remains strong, arguably maintaining its position as the dominant strand. From a very different perspective, Johanna Drucker (2009) argues that the Digital Humanities have been co-opted by a computational perspective inherited from computer science, betraying the humanistic tradition of critical interpretation. Positioning "speculative computing" as the other to the Digital Humanities, she argues that speculative computing "attempts to open the field of discourse to its infinite and peculiar richness as deformative interpretation. How different is it from digital humanities? As different as night from day, text from work, and the force of controlling reason from the pleasures of *delightenment*" (30). Her analysis is less than compelling because it flattens the field's diversity (many working in the Digital Humanities would argue they are practicing what she calls speculative computing), does not attend to critiques of humanities computing from within the field, does not acknowledge work in the second wave, and justifies claims by an idiosyncratic collection of influences. Nevertheless, her critique indicates that the field has not assumed a stable form, even as it puts pressure on traditional practices.

For my purposes, I want to understand the Digital Humanities as broadly as possible, both in its "first wave" practices and "second wave" manifestations (while acknowledging that such classifications are contested). Rather than being drawn into what may appear as partisan infighting, I posit the

Digital Humanities as a diverse field of practices associated with computational techniques and reaching beyond print in its modes of inquiry, research, publication, and dissemination. This perspective is in line with that adopted by the NEH Office of Digital Humanities (Bobley, Rhody, and Serventi, 2011), whose personnel were advised at the program's birth to define the Digital Humanities as broadly as possible. For my purposes, the Digital Humanities include, among other kinds of projects, text encoding and analysis, digital editions of print works, historical research that re-creates classical architecture in virtual reality formats such as *Rome Reborn* and *The Theater of Pompey*, archival and geospatial sites, and, since there is a vibrant conversation between scholarly and creative work in this field, electronic literature and digital art that draws on or remediates humanities traditions.

Scale Matters

Perhaps the single most important issue in effecting transformation is scale. Gregory Crane (2008a) estimates that the upward bound for the number of books anyone can read in a lifetime is twenty-five thousand (assuming one reads a book a day from age fifteen to eighty-five). By contrast, digitized texts that can be searched, analyzed, and correlated by machine algorithms number in the hundreds of thousands (now, with Google Books, a million and more), limited only by ever-increasing processor speed and memory storage. Consequently, machine queries allow questions that would simply be impossible by hand calculation. Timothy Lenoir and Eric Gianella (2011), for example, have devised algorithms to search patents on radio frequency identification (RFID) tags embedded in databases containing six million five hundred thousand patents. Even when hand searches are theoretically possible, the number and kinds of queries one can implement electronically is exponentially greater than would be practical by hand.

To see how scale can change long-established truisms, consider the way in which literary canons typically function within disciplinary practice—in a graduate program that asks students to compile reading lists for the preliminary examination, for example. Most if not all of these works are drawn from the same group of texts that populate anthologies, dominate scholarly conversations, and appear on course syllabi, presumably because these texts are considered to be especially significant, well written, or interesting in other ways. Almost by definition, they are not typical of run-of-the-mill literature. Someone who has read only these texts will likely have a distorted sense of how "ordinary" texts differ from canonized works. By contrast, as Gregory Crane (2008b) observes, machine queries enable one to get a sense of the background conventions against which memorable literary works emerge. Remarkable works endure in part because they complicate, modify, extend, and subvert conventions, rising above the mundane works that surrounded them in their original contexts. Scale changes not only the quantities of texts that can be interrogated but also the contexts and contents of the questions.

Scale also raises questions about one of the most privileged terms in the Traditional Humanities, *reading*. At the level professional scholars perform this activity, reading is so intimately related to meaning that it connotes much more than parsing words; it implies comprehending a text and very often forming a theory about it as well. Franco Moretti (2007:56–57) throws down the gauntlet when he proposes "distant reading" as a mode by which one might begin to speak of a history of *world* literature. Literary history, he suggests, will then become "a patchwork of other people's research, *without a single direct textual reading*" (57; emphasis in original). He continues, "Distant reading: where distance, let me repeat it, is a condition of knowledge: it allows you to focus on units that are much smaller or much larger than the text: devices, themes, tropes—or genres and systems" (57). In this understanding of "reading," interpretation and theorizing are still part of the picture, but they happen not through a direct encounter with a text but rather as a synthetic activity that takes as its raw material the "readings" of others.

If one can perform "distant reading" without perusing a single primary text, then a small step leads to Timothy Lenoir's claim (2008a) that machine algorithms may also count as "reading." Chapter 3 discusses machine reading in more detail, but here I note that from Lenoir's perspective, algorithms read because they avoid what he sees as the principal trap of conventional reading, namely that assumptions already in place filter the material so that one sees only what one expects to see. Of course, algorithms formed from interpretive models may also have this deficiency, for the categories into which they parse units have already been established. This is why Lenoir proclaims, "I am totally against ontologies" (2008a). He points out that his algorithms allow convergences to become visible, without the necessity to know in advance what characterizes them.

Lenoir's claim notwithstanding, algorithms formed from ontologies may also perform the useful function of revealing hitherto unrecognized assumptions. Willard McCarty makes this point about the models and relational databases he used to analyze personification in Ovid's *Metamorphoses*. While the results largely coincided with his sense of how personification works, the divergences brought into view strong new questions about such fundamental terms as "theory" and "explanation" (2005:53–72). As he remarks, "A good model can be fruitful in two ways: either by fulfilling our expectations, and so strengthening its theoretical basis, or by violating them, and so bringing that basis into question" (2008:5).

The controversies around "reading" suggest it is a pivotal term because its various uses are undergirded by different philosophical commitments. At one end of the spectrum, "reading" in the Traditional Humanities connotes sophisticated interpretations achieved through long years of scholarly study and immersion in primary texts. At the other end, "reading" implies a model that backgrounds human interpretation in favor of algorithms employing a minimum of assumptions about what results will prove interesting or important.[4] The first position assumes that human interpretation constitutes the primary starting point, the other that human interpretation misleads and should be brought in after machines have "read" the material. In the middle are algorithms that model one's understanding but nevertheless turn up a small percentage of unexpected instances, as in McCarty's example. Here human interpretation provides the starting point but may be modified by machine reading. Still another position is staked out by Moretti's way of unsettling conventional assumptions by synthesizing critical works that are themselves already synthetic (2000, 2007). Human interpretation remains primary but is nevertheless wrenched out of its customary grooves by the scale at which "distant reading" occurs. Significantly, Moretti not only brackets but actively eschews the level on which interpretation typically focuses, that is, paragraphs and sentences (2007:57).

The further one goes along the spectrum that ends with "machine reading," the more one implicitly accepts the belief that large-scale multicausal events are caused by confluences that include a multitude of forces interacting simultaneously, many of which are nonhuman. One may observe that humans are notoriously egocentric, commonly perceiving themselves and their actions as the primary movers of events. If this egocentric view were accurate, it would make sense that human interpretation should rightly be primary in analyzing how events originate and develop. If events occur at a magnitude far exceeding individual actors and far surpassing the ability of humans to absorb the relevant information, however, "machine reading" might be a first pass toward making visible patterns that human reading could then interpret.

In any case, human interpretation necessarily comes into play at some point, for humans create the programs, implement them, and interpret the results. As Eyal Amiran (2009) observes, the motors driving the process are human desire and interest, qualities foreign to machines. Nevertheless, a human interpreting machine outputs constitutes a significantly different knowledge formation than the Traditional Humanities' customary practice of an unaided human brain-body reading books and arriving at conclusions. Given that human sense-making must necessarily be part of the process, at what points and in what ways interpretation enters are consequential in determining assumptions, methods, and goals. Also at work here is the self-catalyzing dynamic of digital information. The more we use computers, the more we need the large-scale analyses they enable to cope with enormous data sets, and the more we need them, the more inclined we are to use them to make yet more data accessible and machine-readable.

That large-scale events are multicausal is scarcely news, but analysis of them as such was simply not possible until machines were developed capable of creating models, simulations, and correlations that play out (or make visible) the complex interactions dynamically creating and re-creating systems.[5] In turn, the use of tools unsettles traditional assumptions embedded in techniques such as narrative history, a form that necessarily disciplines an unruly mass of conflicting forces and chaotic developments to linear storytelling, which in turn is deeply entwined with the development and dissemination of the codex book. As Alan Liu (2008a) aptly observes about digital technologies (equally true of print), "These are not just tools but tools that we think through." The troops march together: tools with ideas, modeling assumptions with presuppositions about the nature of events, the meaning of "reading" with the place of the human.

The unsettling implications of "machine reading" can be construed as pointing toward a posthuman mode of scholarship in which human interpretation takes a backseat to algorithmic processes. Todd Presner (2008), creator of *Hypermedia Berlin* (2006) and codirector of the *HyperCities* project, reacted strongly when I asked him if digital methods could therefore be seen as erasing the human. As he pointed out, "human" is not a fixed concept but a construction constantly under challenge and revision. Although he conceded that one might characterize certain aspects of the Digital Humanities as posthuman, he insisted the shift should be understood contextually as part of a long history of the "human" adapting to new technological possibilities and affordances. Technologically enabled transformations are nothing new, he argued. Indeed, a major theme in this book is the coevolutionary spiral in which humans and tools are continuously modifying each other (for further elaboration, see Deacon [1998]), Stiegler [1998], Hansen [2006a], and chapters 3, 4, and 5).

The tension between algorithmic analysis and hermeneutic close reading should not be overstated. Very often the relationship is configured not so much as an opposition but as a synergistic interaction. Matthew Kirschenbaum (2009) made this point when discussing a data-mining project designed to rank the letters Emily Dickinson wrote to Susan Huntington Dickinson in terms of erotic language. In interpreting the results, Kirschenbaum and his colleagues sought to understand them by reverse-engineering the sorting process, going back to specific letters to reread them in an attempt to comprehend what kind of language gave rise to a given ranking. The reading practices consisted of what Kirschenbaum calls "rapid shuttling" (2009) between quantitative information and hermeneutic close reading. Rather than one threatening the other, the scope of each was deepened and enriched by juxtaposing it with the other.

The possibility of creating synergistically recursive interactions between close reading and quantitative analyses is also what Stephen Ramsay (2008a) has in mind when he calls for "algorithmic criticism," where the latter word implies hermeneutic interpretation. Positioning himself against a mode of inquiry that praises computer analyses for their objectivity, Ramsay argues that this "scientistic" view (2008b) forsakes the rich traditions of humanistic inquiry that have developed sophisticated and nuanced appreciation for ambiguities. "Why in the world would we want the computer to settle questions?" he asks, proposing instead that computers should be used to open up new lines of inquiry and new theoretical possibilities.

Productive/Critical Theory

What might be these theoretical possibilities? Conditioned by several decades of post-structuralism, many humanistic disciplines associate "theory" with the close scrutiny of individual texts that uncovers and destabilizes the founding dichotomies generating the text's dynamics. A different kind of theory emerges when the focus shifts to the digital tools used to analyze texts and convey results. Jay David Bolter (2008) suggests the possibility of "productive theory," which he envisions as a "codified set of practices." (We may perhaps consider the work of Diane Gromola and Bolter [2003] as characteristic of productive theory.) The ideal, Bolter suggests (2008), would be an alliance (or perhaps integration) of productive theory with the insights won by poststructuralist theories to create a hybrid set of approaches combining political, rhetorical, and cultural critique with the indigenous practices of digital media. Alan Liu (2004) articulates a similar vision when he calls for an alliance between the "cool" (those who specialize in design, graphics, and other fields within digital commercial and artistic realms) and humanities scholars, who can benefit from the "cool" understanding of contemporary digital practices while also enhancing it with historical depth and rich contextualization.

If humanities scholars and the "cool" can interact synergistically, so too can digital media and print. Todd S. Presner speaks of digitality's influence on his print book, *Mobile Modernity: Germany, Jews, Trains* (2007), specifically its network structure. He wanted the book to be experienced as a journey that takes advantage of serendipitous branching to proceed along multiple intersecting pathways. Appropriately for his topic, he envisioned the structure as stations marking intersection points or signaling new directions. In the end, he says, the stations became chapters, but the original design nevertheless deeply informs the work. Matthew Kirschenbaum's print book *Mechanisms: New Media and the Forensic Imagination* (2008) exemplifies traffic in the other direction, from the bibliographic methods developed over the long history of print back into digital media. The idea is to bring to digital media the same materialist emphasis of bibliographic study, using microscopic (and occasionally even nanoscale) examination of digital objects and codes to understand their histories, contexts, and transmission pathways. The result, in Kirschenbaum's phrase, is the emerging field of "digital forensics." Digital networks influence print books, and print traditions inform the ways in which the materiality of digital objects is understood and theorized. Thus two dynamics are at work: one in which the Digital Humanities are moving forward to open up new areas of exploration, and another in which they are engaged in a recursive feedback loop with the Traditional Humanities.

The effects of such feedback loops can be powerfully transformative, as shown in the work of historian Philip J. Ethington, a pioneer in incorporating spatial and temporal data into library records (Hunt and Ethington 1997). For more than a decade, Ethington has undertaken an intellectual journey toward what he calls "the cartographic imagination."[6] Beginning with the insight that spatial and temporal markers are crucial components of any data record, he conceived a number of digital projects in which meaning is built not according to a linear chain of A following B (a form typical of narrative history) but according to large numbers of connections between two or more networks layered onto one another. He writes in the influential essay "Los Angeles and the Problem of Urban Historical Knowledge" that the "key element . . . is a space-time phenomenology, wherein we take historical knowledge in its material presence as an artifact, and map that present through indices of correlation within the dense network of institutions, which themselves are mappable" (2000).

A metaphor may be helpful in understanding this paradigm shift. Just as Ferdinand de Saussure ([1916] 1983) proclaimed that significance is not created by a linear relationship between sign and referent but rather through networks of signifiers, so the movement here is from linear temporal causality to spatialized grids extending in all directions and incorporating rich connections within themselves as well as cross-connections with other grids. The extra dimensions and movements possible in spatial representations compared to linear temporality are crucial in opening up the cartographic imagination to multifocal, multicausal, and nonnarrative modes of historical representation. In a similar vein, Ethington (2007) has argued that history is not primarily a record of what takes place in time but rather what happens in places and spaces. His print book project, a *global* history of Los Angeles, uses these conceptions to create nonnarrative series of networked correspondences instantiated in ten-inch-by-ten-inch format (allowing for twenty-inch page spreads) that incorporates many different kinds of information into temporally marked geospatial grids. Chapter 6 interrogates these issues in more depth by exploring the issues raised by spatial history projects and comparing them with possibilities for progressive, nonteleological and noncolonialist assumptions.

Although the interactions between print and digital media may be synergistic, as in the examples above, they can also generate friction when the Digital Humanities move in directions foreign to the Traditional Humanities. As scale grows exponentially larger, visualization tools become increasingly necessary. Machine queries frequently yield masses of information that are incomprehensible when presented as tables or databases of results. Visualization helps sort the information and make patterns visible. Once the patterns can be discerned, the work of interpretation can begin. Here disagreement among my respondents surfaces, in a debate similar to the controversy over reading. Some argue that the discovery of patterns is sufficient, without the necessity to link them to meaning. Timothy Lenoir's observation (2008a) forcefully articulates this idea: "Forget meaning," he proclaims. "Follow the datastreams." Others, like Stephen Ramsay (2008), argue that data must lead to meaning for them to be significant. If the Digital Humanities cannot do this, Ramsay declares (2009b), "then I want nothing to do with it." The issue is central, for it concerns how the Digital Humanities should be articulated with the Traditional Humanities.

The kinds of articulation that emerge have strong implications for the future: will the Digital Humanities become a separate field whose interests are increasingly remote from the Traditional Humanities, or will it on the contrary become so deeply entwined with questions of hermeneutic interpretation that no self-respecting traditional scholar could remain ignorant of its results? If the Digital Humanities were to spin off into an entirely separate field, the future trajectory of the Traditional Humanities would be affected as well. Obviously, this is a political as well as an intellectual issue. In the case of radical divergence, one might expect turf battles, competition for funding, changing disciplinary boundaries, and shifting academic prestige.

Collaboration

Fortunately, there are strong countervailing tendencies, one of which is collaboration. Given the predominance of machine queries and the size of projects in the Digital Humanities, collaboration is the rule rather than the exception, a point made by John Unsworth when he writes about "the process of shifting from a cooperative to a collaborative model" (2003). Examples of this shift are humanities laboratories in which teams of researchers collectively conceptualize, implement, and disseminate their research. The Humanities Lab at Stanford University, formerly directed by Jeffrey Schnapp, modeled itself on "Big Science," initiating projects that Schnapp calls "Big Humanities" (2009). Implementing such projects requires diverse skills, including traditional scholarship as well as programming, graphic design, interface engineering, sonic art, and other humanistic, artistic, and technical skills. Almost no one possesses all of these skills, so collaboration becomes a necessity; in addition, the sheer amount of work required makes sole authorship of a large project difficult if not impossible.

The program officers at the NEH Office of Digital Humanities confirmed they are seeing more collaborative teams applying for grants (Bobley, Rhody, and Serventi 2011). Indeed, their "Digging into Data" initiative, undertaken with an international group of seven other funding agencies, requires interdisciplinary and international collaborations, not only among humanists but also between computer scientists, quantitative social scientists, and humanists. Brett Bobley emphasized the importance of collaborations that take into account what is happening in other countries, as well as the intellectual synergies created when computer scientist and humanists work together (Bobley, Rhody, and Serventi 2011). He spoke about the excitement that such collaborations can generate, as well as the anxiety about reaching beyond one's comfort zone. "It's OK if participants are uncomfortable," Jennifer Serventi observed, commenting that the momentary discomfort is more than offset by forming expanding networks of colleagues to whom one can turn for expert advice and help (Bobley, Rhody, and Serventi 2011).

Unlike older (and increasingly untenable) practices where a humanities scholar conceives a project and then turns it over to a technical person to implement (usually with a power differential between the two), these collaborations "go deep," as Tara McPherson (2008) comments on the work that has emerged from the online multimodal journal *Vectors*. Conceptualization is intimately tied in with implementation, design decisions often have theoretical consequences, algorithms embody reasoning, and navigation carries interpretive weight, so the humanities scholar, graphic designer, and programmer work best when they are in continuous and respectful communication with one another.

As a consequence of requiring a clear infrastructure within which diverse kinds of contributions can be made, "Big Humanities" projects make possible meaningful contributions from students, even as undergraduates. As I write these words, thousands of undergraduates across the country are engaged in writing essays that only their teachers will see—essays that will have no life once the course ends. As Jeffrey Schnapp (2009) and Gregory Crane (2008a) note, however, students can complete smaller parts of a larger web project, ranging from encoding metadata to implementing more complex functionalities, that continues to make scholarly contributions long after they have graduated. In Timothy Lenoir's *Virtual Peace* project (Lenoir et al. 2008), undergraduates did much of the virtual reality encoding, and in Todd Presner's *Hypermedia Berlin* (2006) and *HyperCities* (Presner et al. 2008), Barbara Hui and David Shepard, both graduate students at the time, were crucial members of the project team. Mark Amerika has instituted a similar practice at the University of Colorado, supervising undergraduate contributions to *Alt-X Online Network* (Amerika n.d.), a large database that continues to grow through generations of students, becoming richer and more extensive as time goes on. Eric Rabkin (2006), among others, writes about using digital platforms to give students enhanced literacy skills, including style, collaboration, and a sense of audience. He notes that encouraging students to incorporate graphics, hyperlinks, and so on in their work "makes them exquisitely aware that . . . this technology is more than just an extension, as heels make us a tad taller, but rather a transformative reality, like the automobile" (142). Brett Bobley stresses the importance of reaching out to students when he pointed out that the summer Institutes for Advanced Studies funded by the NEH Office of Digital Humanities specifically invites graduate students to apply. The Office of Digital Humanities sees this as a crucial part of "growing the field" of the Digital Humanities (Bobley, Rhody, and Serventi 2011).

Collaboration may also leap across academic walls. For the first time in human history, worldwide collaborations can arise between expert scholars and expert amateurs. The latter term, I want to insist, is not an oxymoron. The engineer who spends his evenings reading about the Civil War, the accountant who knows everything about Prussian army uniforms, the programmer who has extensively studied telegraph code books and collected them for years—these people acquire their knowledge not as professionals practicing in the humanities but as private citizens with passion for their subjects. I recall hearing a historian speak with disdain about "history buffs." Like history, most professional fields in the humanities have their shadow fields, for example, people who want to argue that Shakespeare did not write the works for which he is credited. Scholars often regard such activity as a nuisance because it is not concerned with questions the scholar regards as important or significant. But this need not be the case. Working together within a shared framework of assumptions, expert scholars and expert amateurs can build databases accessible to all and enriched with content beyond what the scholars can contribute.

An example is the *Clergy of the Church of England Database* directed by Arthur Burns (Clergy of the Church n.d.), in which volunteers collected data and, using laptops and software provided by the project, entered them into a database. *Hypermedia Berlin* (Presner 2006) offers another model by providing open source software through which community people can contribute their narratives, images, and memories, while *HyperCities* (Presner et al. 2008) invites scholars and citizens across the globe to create data repositories specific to their regional histories. In addition to contributions to scholarship, such projects would create new networks between scholars and amateurs, from which may emerge, on both sides of the disciplinary boundary, renewed respect for the other. This kind of model could significantly improve the standing of the humanities with the general public.

Collaboration is not, however, without its own problems and challenges, as scientific research practices have demonstrated. Aside from questions about how collaborative work will be reviewed for tenure and promotion, internal procedures for distributing authority, making editorial decisions, and apportioning credit (an especially crucial issue for graduate students and junior faculty) are typically worked out on a case-by-case basis in Digital Humanities projects. So too are questions of access and possibilities for collaborations across inside/outside boundaries, such as deciding whether the XML (extensible markup language) metadata will be searchable or downloadable by users, and whether search algorithms can be modified by users to suit their specific needs. Discussing these questions in the context of the *Walt Whitman Archive*, Matthew Cohen (2011) stresses the importance of an "ethics of collaboration." Precedents worked out for scientific laboratories may not be appropriate for the Digital Humanities. While the lead scientist customarily receives authorship credit for all publications emerging from his laboratory, the Digital Humanities, with a stronger tradition of single authorship, may choose to craft very different kinds of protocols, including giving authorship credit (as opposed to acknowledgement) for the creative work of paid technical staff.

At the same time, as collaborative work becomes more common throughout the Digital Humanities, tenure and promotion committees will need to develop guidelines and criteria for evaluating collaborative work and digital projects published online. Indeed, Brett Bobley (Bobley, Rhody, and Serventi 2011) identified tenure and promotion issues as one of the most important challenges the Digital Humanities faces in moving forward. The Modern Language Association has tried to respond by issuing guidelines for evaluating digital projects, but recently James P. Purdy and Joyce Walker (2010) urge instead the development of joint criteria that apply both to print and digital media, arguing that separate guidelines reinforce the print/digital binary. Among the rubrics they discuss is "Design and Delivery," a category that requires thoughtful use of print design as well as careful consideration of the best mode of dissemination. Their approach emphasizes flexible coordination between print and digital media, as well as an implicit recognition that print is no longer the default medium of communication in the humanities.

Databases

While scale and collaboration transform the conditions under which research is produced, digital tools affect research both at the macro level of conceptualization and the micro level of fashioning individual sentences and paragraphs. David Lloyd (2008), a scholar working in Irish literature at the University of Southern California, recounted how he worked with his print essay on Irish mobility in the nineteenth century to reenvision it for digital publication in *Vectors*. Working with the flexible database form that Tara McPherson and her coeditor Steven Anderson devised, Lloyd rewrote his text, removing all the structures of coordination and subordination. The fragments were then entered into the database in a recursive process of specifying categories and modifying them as the work proceeded. Lloyd noted that in cutting out subordination and coordination, something was lost—namely the coherence of his argument and crafted prose of his original essay, a loss he felt acutely when he was in the midst of the fragmenting process. But something was gained as well. The effect of the database format, Lloyd said, was to liberate contradictory and refractory threads in the material from the demands of a historically based argument, where they were necessarily smoothed over in the interest of coherence. By contrast, database elements can be combined in many different ways, depending on how a reader wants to navigate the interface. Lloyd along with designer Erik Loyer (Lloyd and Loyer 2006) visualized the topics as potatoes in a field that the reader navigates by "digging" them. The result, Lloyd suggested, was both a richer context and a challenge to the reader to understand their interactions. Like much electronic work, the task requires more patience and work on the reader's part than does a traditional linear narrative, with the payoff being an enhanced, subtler, and richer sense of the topic's complexities. Lloyd, readily acknowledging that some research is no doubt best presented in print, was nevertheless sufficiently impressed with the advantages of a database structure to consider using it for his future poetic writing.

Another advantage of databases is the ability to craft different kinds of interfaces, depending on what users are likely to find useful or scholars want to convey. Given a sufficiently flexible structure, a large archive can have elements coded into a database for which different scholars can construct multiple interfaces. As Tara McPherson points out (2008), the same repository of data elements can thus serve different purposes to different communities. Teams of collaborators might work together to create a shared database, with each team creating the interface best suited for its research purposes. Thus each team's efforts are leveraged by the magnitude of the whole, while still preserving the priorities of its own needs and criteria. Another kind of example is Kimberly Christen's *Mukurtu: Wumpurrarini-kari* website on Australian aboriginal artifacts, histories, and images. She provided aboriginal users with a different interface offering more extensive access than the general public, giving different functionalities to each group (see Christen

2009a and 2008). In addition, the website honors tribal cultural practices, such as the prohibition on viewing images of deceased persons. When such a person appears in an archived photograph, the indigenous user is first warned and then asked if she nevertheless wishes the view the image. Rejecting the mantra "Information wants to be free," Christen suggests an alternative: "Information wants to be responsible" (2009b).

The collaborations that databases make possible extend to new kinds of relationships between a project's designer and her interlocutors. Sharon Daniel (2008), discussing her work with drug addicts and women incarcerated in California prisons, declared that she has moved away from an emphasis on representation to participation. She sees her digital art work, her award-winning *Public Secrets*, for example, as generating context "that allows others to provide their own representation," particularly disenfranchised communities that might otherwise not have the resources to create self-representations. Eschewing documentary forms that emphasize a single authorial perspective, Daniel created a database structure that allows her interlocutors to speak for themselves. Because of her political commitment to participation, the database structure is crucial. Daniel's method (similar in its procedures to many digital projects, like the database created by Lloyd and Loyer), is to locate the topic's central problematics and design the data structure around them. With *Blood Sugar*, a companion piece to *Public Secrets*, the fact that addiction is both biological and sociological provided the essential parameters.

The emphasis on databases in Digital Humanities projects shifts the emphasis from argumentation—a rhetorical form that historically has foregrounded context, crafted prose, logical relationships, and audience response—to data elements embedded in forms in which the structure and parameters embody significant implications. Willeke Wendrich, director of the digital *Encyclopedia of Egyptology*, spoke eloquently about the ethical significance of this shift.[7] Working in archeology, a field in which researchers sometimes hoarded artifacts and refused access to them to aggrandize their personal power base, Wendrich argues that database forms and web dissemination mechanisms allow for increased diversity of interpretation and richness of insights, because now the data are freely available to anyone. Of course, those who design such websites still influence the range and direction of interpretation through selections of material, parameters chosen for the database structures, and possible search queries. As Geoffrey Bowker and Susan Leigh Star (2000) have persuasively argued, the ordering of information is never neutral or value-free. Databases are not necessarily more objective than arguments, but they are different kinds of cultural forms, embodying different cognitive, technical, psychological, and artistic modalities and offering different ways to instantiate concepts, structure experience, and embody values (Vesna 2007; Manovich 2002).

Multimodal Scholarship

In addition to database structures and collaborative teams, the Digital Humanities also make use of a full range of visual images, graphics, animations, and other digital effects. In best-practice projects, these have emotional force as well conceptual coherence. Caren Kaplan (2008) spoke to this aspect of her project *Dead Reckoning*, developed for *Vectors* in collaboration with designer Raegan Kelly (Kaplan with Kelly 2007). After encountering a wealth of cultural analysis and technical information about aerial surveillance and targeting, the user is presented with a section in which she can manipulate the target image herself. When it centers over Hiroshima, the emotional impact of occupying the position of the (virtual) bomber creates a strong affective resonance. Alice Gambrell and Raegan Kelly's *Stolen Time Archive* (Gambrell with Kelly 2005), a collection of female office worker ephemera from the 1940s and 1950s and later zines, achieves a different kind of emotional impact through the ambiguity of "stolen time." From the employer's point of view, time theft occurs when an employee produces such objects as the zines; from the worker's viewpoint, time is stolen from her by an alienating capitalist system, represented in the archive through posters and advice manuals intended to refashion her subjectivity so it will be more malleable for the system. From the user's point of view, time spent perusing the archive and contemplating its significance is the productive/unproductive dynamic revealing ambiguities at the heart of the archive. For these and similar works, multimodality and interactivity are not cosmetic enhancements but integral parts of their conceptualization.

In light of such developments, Timothy Lenoir (2008a) draws the conclusion that the Digital Humanities' central focus should be on developing, incorporating, and creating the media appropriate for their projects. "We make media," Lenoir proclaims. "That's what we do." A case in point is the peace and conflict simulation *Virtual Peace: Turning Swords to Ploughshares* (Lenoir et al. 2008) that he and his collaborators created. The collaboration involved professors, students, and programmers from the Virtual Heroes commercial game company. The simulation runs on Epic Games' Unreal Tournament game engine, for which Virtual Heroes had a license and adapted with tools, scripts, and other assets. Instead of preparing troops for war (as do many military simulations), this project aims to improve conflict resolution skills of stakeholders responding to an emergency (the simulation makes extensive use of the data from Hurricane Mitch in 1998, which caused extensive damage in Honduras and other places). The project was funded by a $250,000 MacArthur grant; the inclusion of commercial programmers indicates that large projects such as this require outside funding, either from corporate sponsors or foundations and granting agencies. Traditional humanities scholars, accustomed to requiring nothing more than a networked computer and some software, sometimes critique projects like *Virtual Peace* and *HyperCities* because they rely on commercial interests (*HyperCities* makes extensive use of Google Maps and Google Earth). Presner (2008) remarked that he has been told that he is "in bed with the devil."[8]

The remark points to tensions between theoretical critique and productive theory. In poststructuralist critique, a hermeneutic of suspicion reigns toward capitalism and corporations, while in the Digital Humanities, a willingness prevails to reach out to funders (sometimes including commercial interests). Cathy N. Davidson and David Theo Goldberg (2004) suggest we should move past the hermeneutic of suspicion: "What part of our inability to command attention is rooted in humanists' touting of critique rather than contribution as the primary outcome of their work? . . . Is it not time we critiqued the mantra of critique?" (45). Some scholars in the Digital Humanities, including Presner (2008) and Anne Balsamo, are already moving in this direction. As Balsamo argues in *Designing Culture: The Technological Imagination at Work* (2011), humanities scholars should seize the initiative and become involved in helping to develop the tools our profession needs. We cannot wait, Balsamo contends, until the tools arrive readymade (and often ill-made for our purposes). Rather, we should get in on the ground floor through collaborations not only among ourselves (as in the Project Bamboo Digital Humanities Initiative) but also with commercial companies such as Google.

Code

Another area of tension between poststructuralist approaches and productive theory is the environment in which Digital Humanities work. Underlying machine queries, database structures, and interface design is a major assumption that characterizes the Digital Humanities as a whole: that human cognition is collaborating with machine cognition to extend its scope, power, and flexibility. The situation requires both partners in the collaboration to structure their communications so as to be legible to the other. For humans, this means writing executable code that ultimately will be translated into a binary system of voltages; for the machine, it means a "tower of languages" (Cayley 2002; Raley 2006) mediating between binary code and the displays the user sees. Multiple implications emerge from this simple fact. If the transition from handwriting to typewriting introduced a tectonic shift in discourse networks, as Friedrich Kittler (1992) has argued, the coupling of human intuition and machine logic leads to specificities quite different in their effects from those mobilized by print. On the human side, the requirement to write executable code means that every command must be explicitly stated in the proper form. One must therefore be very clear about what one wants the machine to do. For Tanya Clement (2008a), a graduate student at the University of Maryland working on a digital analysis of Gertrude Stein's *The Making of Americans*, this amounts in her evocative phrase to an "exteriorization of desire." Needing to translate desire into the explicitness of unforgiving code allows implications to be brought to light, examined, and modified in ways that may not happen with print. At the same time, the nebulous nature of desire also points to the differences between an abstract computational model and the noise of a world too full of ambiguities and complexities ever to be captured fully in a model.

The necessity for executable code creates new requirements for digital literacy. Not every scholar in the Digital Humanities needs to be an expert programmer, but to produce high-quality work, scholars certainly need to know how to talk to those who are programmers. The Digital Humanities scholar is apt to think along two parallel tracks at once: what the surface display should be, and what kinds of executable code are necessary to bring it about. This puts subtle pressure on the writing process, which in turn also interacts with the coding. Reminiscent of David Lloyd's excision of coordination and subordination, many writers who move from print to digital publication notice that their writing style changes. In general, the movement seems to be toward smaller blocks of prose, with an eye toward what can be seen on the screen without scrolling down and toward short conceptual blocks that can be rearranged in different patterns. The effects spill over into print. Alexander R. Galloway and Eugene Thacker's *The Exploit: A Theory of Networks* (2007), a print text about digital networks, parses the argument in part through statements in italics followed by short explanatory prose blocks, so that the book can be read as a series of major assertions (by skipping the explanations), short forays into various questions (by picking and choosing among blocks), or straight through in traditional print reading fashion.

Given the double demand for expertise in a humanistic field of inquiry and in computer languages and protocols, many scholars feel under pressure and wonder if they are up to the task. Even talented scholars recognized as leaders in their fields can occasionally have doubts. Rita Raley (2008), pointing out that she is trained in archival research and not in computer programming, wondered if, in writing about code poetry, she is committing herself to a field in which she is not a master. Tanya Clement (2008a), whose work as a graduate student has already achieved international recognition, says that she is "not a Stein scholar" and is consequently hesitant about presenting her quantitative analyses of *The Making of Americans* (2008b) to Stein scholars. (I should add that these scholars are exemplary practitioners producing cutting-edge scholarship; their doubts reveal more about the field's problematics than any personal deficiencies.) The problems become explicit when double expertise is formalized into an academic curriculum, such as in the computational media major recently instituted at Georgia Tech. Ian Bogost (2009), one of the faculty members leading the program, spoke eloquently about the difficulties of forging requirements fully responsive both to the demands of the Computer Science Department and to the expectations of a humanities major. I suspect there is no easy solution to these difficulties, especially in this transitional time when dual expertise is the exception rather than the rule. In the future, academic programs such as Georgia Tech's computational media and the humanities computing majors at King's College may produce scholars fluent both in code and the Traditional Humanities. In the meantime, many scholars working in the field are self-taught, while others extend their reach through close and deep collaborations with technical staff and professionals in design, programming, etc.

Future Trajectories

I asked my respondents what percentages of scholars in the humanities are seriously engaged with digital technologies. Many pointed out that in a sense, virtually everyone in the humanities is engaged with digital technologies through e-mail, Google searches, web surfing, and so on. But if we take "seriously" to mean engagements that go further into web authoring and the construction of research projects using digital tools, the percentages were generally low, especially if averaged over the humanities as a whole. In September 2005, participants in the Summit on Digital Tools in the Humanities at the University of Virginia estimated that "only about six percent of humanist scholars go beyond general purpose information technology and use digital resources and more complex digital tools in their scholarship" (*Summit* 2005:4). Given developments since then, my estimate of where we are currently is about 10 percent. But this figure may be misleading, for as my interviewees agreed, the numbers are generationally skewed, rising quickly within the younger ranks of the professoriate and even more so among graduate students. Many people estimated 40–50 percent of younger scholars are seriously engaged. This demographic suggests that involvement in the Digital Humanities will continue to increase in the coming years, perhaps hitting about 50 percent of humanists when those who are now assistant professors become full professors in ten to fifteen years. This prediction suggests that the scholarly monograph will not continue indefinitely to be the only gold standard and that web publishing will not only be commonplace but will attain equal standing with print.

It would be naïve to think that this boundary-breaking trajectory will happen without contestation. Moreover, practitioners in the field recall similar optimistic projections from fifteen or twenty years ago; in this respect, prognostications for rapid change have cried wolf all too often. Among those skeptical that progress will be swift are Eyal Amiran (2009), cofounder of *Postmodern Culture*, one of the first scholarly journals to go online, and Jay David Bolter (2008), who remarked that literature departments in particular seem "unreasonably resistant" to introducing digital technologies into the humanities. Nevertheless, new factors suggest a critical mass has been reached. Foremost is the establishment of robust Digital Humanities centers at the University of Maryland; King's College London; the University of Nebraska; the University of Texas; the University of California, Santa Barbara; the University of California, Los Angeles; and many other institutions. Brett Bobley (Bobley, Rhody and Serventi 2011) at the NEH reported that the number of inquiries the Office of Digital Humanities receives from scholars wanting to start Digital Humanities centers is increasing, a trend confirmed by the existence of such centers at institutions such as the University of Maryland and the University of Nebraska.

A concurrent development is the marked increase in the number of scholarly programs offering majors, graduate degrees, and certificate programs in the Digital Humanities, with a corresponding growth in the numbers of students involved in the field. Willard McCarty (2009) extrapolates from this development to see a future in which humanities scholars are also fluent in code and can "actually make things." Once critical mass is achieved, developments at any one place have catalyzing effects on the field as a whole. Intimately related to institutionalization and curricular development are changing concepts and presuppositions. The issues discussed here—scale, productive/critical theory, collaboration, databases, multimodal scholarship and code—are affecting the structures through which knowledge is created, contextualized, stored, accessed, and disseminated.

Among my interviewees, scholars with administrative responsibilities for program development typically had thought most about future trajectories and were most emphatic about the transformative potential of digital technologies. Three examples illustrate this potential. Kenneth Knoespel, chair of the School of Literature, Culture and Communication (LCC) at Georgia Tech, pointed to the cooperative ventures his faculty had underway with the engineering and computer science departments; in these alliances, humanities students provided valuable input by contextualizing technical problems with deep understandings of social, cultural, and historical embeddings (Knoespel 2009). In his view, digital media provide a common ground on which humanities scholars can use their special skills in interpretation, critical theory, close reading, and cultural studies to enhance and codirect projects with their colleagues in the sciences, engineering, and social sciences. With team-based projects, sophisticated infrastructure, tech-savvy faculty and an emphasis on studio work, LCC has reimagined itself as a program that combines textual analysis with a wide variety of other modalities (Balsamo 2000). In Knoespel's view, LCC is about reenvisioning not only the humanities but higher education in general (Knoespel 2009).

A second example is the Transcriptions initiative that Alan Liu, chair of the English Department at the University of California, Santa Barbara, has spearheaded. Like Knoespel, Liu actively reaches out to colleagues in engineering and scientific fields, using digital technologies as a common ground for discussions, projects, and grant applications. He sees this as a way to reenvision and reinvigorate humanities research along dramatically different lines than Traditional Humanities, while also providing support (financial, administrative, and institutional) for text-based research as well. Tara McPherson, coeditor of *Vectors* at the University of Southern California, also believes that humanistic engagements with digital technologies have the potential to reimagine higher education. Like Knoespel and Liu, she takes an activist approach, finding the funding, infrastructure, and technical support to help humanities scholars move into multimodal projects and envision their research in new ways. When faculty move in these directions, more than their research is affected; also reimagined is their pedagogy as their students become collaborators in Digital Humanities projects, their relationships with colleagues outside the humanities, and their vision of what higher education can achieve and contribute.

Two Strategies for the Digital Humanities: Assimilation and Distinction

As mentioned earlier, I made site visits to the CCH at King's College London and the LCC at Georgia Tech to supplement information gathered in the interviews. Whereas the interviews identified major themes that distinguish the digital from the Traditional Humanities, the site visits suggested the importance of institutional structures for thinking about how the Digital Humanities can be articulated together with more traditional approaches. Two major strategies became apparent in these visits: assimilation and distinction.

Assimilation extends existing scholarship into the digital realm; it offers more affordances than print for access, queries, and dissemination; it often adopts an attitude of reassurance rather than confrontation. Distinction, by contrast, emphasizes new methodologies, new kinds of research questions, and the emergence of entirely new fields. Assimilation strategies are pursued, for example, by *Postmodern Culture*, electronic editions of print texts, and Willard McCarty's fine book *Humanities Computing* (2005). The distinction side might list *Vectors*, much digital fiction and poetry, and Timothy Lenoir and colleagues' *Virtual Peace* simulation. Both strategies have the potential to transform, but they do so by positioning themselves differently in relation to humanities research, formulating research questions in different ways, pursuing different modes of institutionalization, and following different kinds of funding strategies. Comparing and contrasting their respective strengths and limitations enables a more nuanced view of how the Digital Humanities may be articulated with the Traditional Humanities and how this positioning entails a host of other considerations, including credentialing, evaluation, rhetorical posture, and institutional specificity.

As if on cue, London provided a fine light drizzle for my morning walk to King's College. Courtesy of my host Willard McCarty, I am meeting with Harold Short, director of the CCH, and later holding a group interview with a dozen or so scholars who have mounted a variety of digital projects. CCH arguably boasts the most extensive curricula and the most developed program of Digital Humanities in the world. I want to understand how this robust program was built and why it continues to flourish.

CCH began with an undergraduate teaching major and, under the expert guidance of Short, expanded into a major research unit as well as a program offering a master's and a PhD in the Digital Humanities. In addition to five full-time faculty and one emeritus, CCH now employs between thirty and forty academic-related staff, including programmers and designers. About half are supported by grants and other soft money; the rest are permanent staff positions. Each staff member is typically assigned to multiple projects at a time. Although some of these researchers initiate projects, they often develop them through collaboration with other humanities scholars. An anecdote Short (McCarty and Short 2009) tells about one of his colleagues, David Carpenter, suggests a typical pattern of development. Carpenter came to Short because he had been told by a colleague that he would not get funding for his project unless there was an electronic component. He wanted to do an edition of the fine rolls compiled during Henry III's reign, lists of fines people paid if they wanted to marry, if they preferred to remain unmarried, etc. During their consultation, it became apparent that if grant money was paid to CCH, Carpenter would have less money for his research assistants. At that point, as Short put it, "he got serious." If real money was at stake, the electronic component would have to be more than window dressing; it would have to have a research payoff. Carpenter returned and (according to Short) asked, "Does that mean that if we use the methods you are proposing, I'll be able to ask to be shown all the fines to markets in North Hampshire between the years 1240 and 1250?" (McCarty and Short 2009). When Short replied in the affirmative, Carpenter began to understand the potential of the digital component and became one of the strongest supporters of CCH.

While it may appear that CCH thus functions as a technical support unit, Short insists that CCH staff are collaborators, researchers in their own right who bring ideas to the table rather than simply implementing the ideas of others. "From an early stage," Short reports, "we looked at collaborative research partners as equals. They felt the same way" (McCarty and Short 2009). Moreover, CCH staff are adept at crafting the collaborations into projects that are candidates for grant money. The record is impressive. According to Short, CCH has generated close to £18 million in research income, £6 million of which has come to CCH, £6 million to King's College, and £6 million to outside partners. "We have never been seen as a threat that will take away funding," Short observes, "because we were bringing in funding" (McCarty and Short 2009). Through teaching, collaborative research, and procuring funding, CCH has made itself integral to the college's normal operations. When asked how he accounts for CCH's success, Short replied, "embeddedness" (McCarty and Short 2009).

The kinds of projects CCH develops were vividly on display in the group interview. Stephen Baxter (McCarty et al. 2009), a medieval historian, spoke eloquently about the digitization of the *Domesday Book*, an eleventh-century historical document that surveyed people, landholdings, and natural resources. Digitizing this important text allowed sophisticated machine analysis that discerned patterns not previously known, a technique generally known as knowledge discovery and data mining. Researchers were able to determine how the survey had been conducted and what circuits were visited in what order; in addition, they were to a large extent able to overcome the problem of homonyms (only first names were given, so the document recorded, for example, three hundred or more Arthurs with no clear way to assign them to particular people). As a result, researchers arrived at a better view of landholders and were able to make new inferences about the structure of contemporary aristocracy. So important has this digital analysis been, Baxter remarked, that no one working in the period can afford to ignore the results.

Another kind of project (mentioned earlier), directed by Arthur Burns, exemplifies the possibilities for collaborations between professional scholars and expert amateurs. "*Clergy of the Church of England Database*" aims to create a relational database of the careers of all Anglican clergy in England and Wales between 1540 and 1835. So far it has gathered half a million records that link person to place, available on the project's website. "It forces a huge change in the way we work," Burns remarks (McCarty et al. 2009), including partnerships with "genealogical historians looking for British ancestors. We set up an interactive query site; we created an academic journal; it's become a crossroads." With fifty regional archives to survey, Burns realized that "we could pay for the mapping but we couldn't pay for the researchers." The ingenious solution was to contract with one hundred volunteers working across the United Kingdom, giving them laptops loaded with the appropriate software. "It has become a collaborative scholarship," Burns remarks, noting that "we ended up being on the cutting edge of digital scholarship, something which we'd never anticipated."

So far, these examples focus on digitizing historical records and documents. They extend the analytical procedures and strategies capable of generating new knowledge, but their scope is necessarily limited to the parameters implicit in preexisting documents. A different model, more toward the distinction end of the spectrum than assimilation, was created by the King's Visualisation Laboratory, a 3-D rendering of how the Theater of Pompey may have appeared. Working with a team of archeologists, Hugh Denard (Denard 2002; McCarty et al. 2009) and his senior colleague Richard Beacham, along with other colleagues, prepared an initial visualization to present to the Rome city authorities to gain permission for an excavation. In part because of the visualization's persuasive force, permission was granted, and a second version incorporated the results of that work to refine and extend their model. Denard (McCarty et al. 2009) suggested that in the future, classicists and archeologists might get their first exposure to the theater through the visualization, in effect shifting the sensorium of knowledge construction from text-based artifacts to an interactive 3-D rendering that allows them to change perspective; zoom in and out of details, floor plans, and architectural features; and imaginatively explore the space to visualize how classical plays might have been performed. Building on this work, Denard and an interdisciplinary international group of colleagues collaborated to create the London Charter for the Use of 3-Dimensional Visualization in the Research and Communication of Cultural Heritage, a document setting out standards for projects of this kind.

In speaking about the amount of data comprised by the Pompey project, Denard (McCarty et al. 2009) commented that "it embodies five monographs [worth of effort and material]." This idea was echoed around the table by Charlotte Roueche, Arthur Burns, Jane Winters, and Stephen Baxter, among others. A common problem many of these researchers encountered was having the complexity, extent, and achievement of their digital projects appropriately recognized and reviewed. This was in part because there are only a few people with the experience in digital research to properly evaluate the projects, and in part because other evaluators, experts in the field but unused to digital projects, did not have an accurate understanding of how much work was involved. Denard (McCarty et al. 2009) commented that "our main [financial] resource is the Research Council. . . . The rule about peer review is that you have to draw your reviewers from within the UK, and often there isn't another person who is qualified in the country." Jane Winters (McCarty et al. 2009) commented, "Nobody conceives [the project] as a whole. Sometimes you are seen as technical project officers rather than researchers or research assistants."

The group also noted the gap between the many that use their projects and the few that want to go on to create digital projects themselves.

Charlotte Roueche (McCarty et al. 2009) commented, "You can't do classics without using of the collections online, so virtually every classicist uses Latin collections. As users, they are quite adjusted. Oddly, it doesn't occur to them that they could be producers." When asked what percentage of researchers in their field create digital resources, Arthur Burns estimated "around 5 percent" (McCarty et al. 2009). Jane Winters concurred, noting that "we produced resources for medievalists and early modernists . . . [but] amateurs tend to be the people with the real engagement, who ask questions about your production. It just doesn't happen with academic colleagues; they are just users. I tend toward 5 percent" (McCarty et al. 2009). Explaining why specialists find the resources invaluable but do not feel moved to create them themselves, Burns explained, "If you went to someone negative [about creating digital resources himself], he'd say it's not the job of historians to produce texts; it's to interpret texts, and what digital scholarship is doing is what archives do . . . it's not our [historians'] job." (McCarty et al. 2009). Roueche commented that "for me to carry out certain analysis, I have to have the texts in digital form. Once they are, I'd be able to formulate new questions. . . . I feel like it is the next generation who will do these things, and our job is to prepare the resources" (McCarty et al. 2009). In the meantime, these scholars inevitably face difficulties in having their labor seen as something other than "donkey work," as Roueche (McCarty et al. 2009) ironically called it.

In summary, then, the assimilative strategy offers the advantage of fruitful collaborations with historically oriented humanities research, where it pays rich dividends in the discovery of implicit patterns in complex data, the accessibility of rare texts to the scholarly community, the ability to formulate new questions that take advantage of machine analyses, and the enlistment of expert amateurs into research activities, especially in contributions to large databases. It has the disadvantages of being underevaluated by print-based fields, being partially invisible in the amount of effort it requires and, in some cases, being seen as a low-prestige activity removed from the "real" work of the Traditional Humanities. Institutionally, assimilation seems to work best when there is a large amount of historical research suitable for digitization, a strong national infrastructure of grant money to support such research, and nonprofit collaborators, often from artistic and library fields, who are willing partners capable of providing money and other resources for the projects.

To explore the advantages and limitations of a distinction strategy, I take as my example the LCC at Georgia Tech, with which I have long-standing ties. From a rather undistinguished English department regarded as a service unit providing composition instruction for technical departments, LCC began over twenty years ago to transform into a vital cutting-edge multidisciplinary school. It was one of the first programs in the country to launch a track in literature and science; cultural studies followed as an important emphasis, and more recently, digital media studies. Under the enlightened leadership of Knoespel, a polymath whose research interests range from digitizing Newton's manuscripts to architectural visualizations (Knoespel 2009), LCC has attracted a distinguished faculty in digital media that includes Jay Bolter, Janet Murray, Ian Bogost, and others. LCC now offers two tracks for the BS: science, technology, and culture; and computational media. The latter, a joint program between the Computer Science Department and LCC, requires an especially high degree of technical proficiency. On the graduate level, LCC offers two tracks for the MS, human computer interaction and digital design, and more recently has added a PhD in digital media. LCC is thus comparable to CCH in the depth of curricula, the full spectrum of degrees, and the strength of the faculty.

LCC's emphasis, however, is quite different than that of CCH. The difference is signaled by its mission statement, which says the school provides "the theoretical *and practical* foundation for careers as digital media researchers in academia *and industry*" (emphasis added). Whereas CCH works mostly with art and cultural institutions, LCC has corporate as well as nonprofit partners and places many of its graduates in for-profit enterprises. Corporate sponsors include Texas Instruments, Warner Brothers, Turner Broadcasting, and Sony, and on the nonprofit side, the American Film Institute, the National Science Foundation, NEH, and the Rockefeller Foundation. To provide the necessary practical expertise, LCC boasts a large number of labs and studios, with courses oriented toward production as well as theory. These include the Mobile Technologies Lab, Experimental Game Lab, eTV Production Group, Synaesthetic Media Lab, MMOG (Massive Multiplayer Online Game) Design and Implementation Lab, and the Imagination, Computation, and Expression Studio, among others. Student projects include tool-building, corporate-oriented projects such as "Virtual IKEA: Principles of Integrating Dynamic 2D Content in a Virtual 3 D Environment," game studies such as "Key and Lock Puzzles in Procedural Gameplay," and artistic and theoretical projects.

Instead of the "embeddedness" characteristic of CCH at King's College, LCC focuses not on digitizing historical resources but rather on contemporary and emerging digital media such as computer games, cell phones, and eTV. While Traditional Humanities research goes on at LCC, there is relatively little overlap with the digital media (with significant exceptions such as Ian Bogost and others). In line with Harold Short's comment that CCH was not perceived as a threat because it brought in valuable grant money to the university, the success of the digital media program at LCC in bringing in grants and finding corporate sponsors has helped to offset the disgruntlement that some faculty in the Traditional Humanities might feel. Nevertheless, there is stronger tension between the two groups than is the case at CCH at King's because digital media is seen more as a separate field than as an integral part of humanities research.

The distinction approach, as it is implemented at LCC and elsewhere, aims to create cutting-edge research and pedagogy specifically in digital media. To a significant degree, it is envisioning the future as it may take shape in a convergence culture in which TV, the web, computer games, cell phones, and other mobile devices are all interlinked and deliver cultural content across as well as within these different media. In contrast to CCH researchers' feeling that their work is used but not properly appreciated, the work of the digital media component at LCC is highly visible nationally and internationally and widely understood to represent state-of-the-art research. On the minus side, the relation of this research to the Traditional Humanities is less clear, more problematic, and generally undertheorized. The advantages and limitations are deeply interlinked. Because there is less connection with the Traditional Humanities, the digital media curricula and research can explore newer and less text-based modalities, but for that very reason, it contributes less to the Traditional Humanities and has weaker ties to them.

The comparison of LCC and CCH suggests that neither strategy, assimilation or distinction, can be judged superior to the other without taking into account the institutional contexts in which digital media programs operate. With a rich tradition of historical research and many faculty interested in the digitization of print and manuscript documents, CCH flourishes because of the many connections it makes with ongoing humanities research. Located within a technical institute without a similarly robust humanities tradition but with strong engineering and computer science departments, the digital media program at LCC succeeds because it can move quickly and flexibly into new media forms and position itself at the frontier of technological innovation and change. Institutional specificity is key when deciding on which strategy may be more effective, more robust, and more able to endure budget cuts, tight finances, and other exigencies that all universities experience from time to time.

I have chosen CCH and LCC for comparison because they represent two ends of a spectrum and thus clearly demonstrate the limitations and advantages of the assimilation and distinction strategies. There are, however, successful hybrid programs representing combinations of the two strategies, among which I include the program in electronic writing at Brown University, the Maryland Institute for Technical Humanities at the University of Maryland, and the Institute for Advanced Technologies at the University of Virginia. Goldsmiths, University of London, strong in the Traditional Humanities, illustrates a hybrid approach in offering two tracks for its MA in digital media: the first in theory, and the second in theory and practice. Other hybrid programs are flourishing at the newly founded DesignLab at the University of Wisconsin, Madison, and in Virginia Commonwealth University's PhD program in media, art and text. The challenge for such programs is to find ways to incorporate the insights of the Traditional Humanities, especially poststructuralist theory and gender, ethnic, and race studies, into practice-based research focusing primarily on the acquisition and deployment of technical skills.[9] Such integration will be key to formulating strong bonds between the Traditional and Digital Humanities.

Although the rubrics discussed above and the institutional issues defy easy summary, their breadth and depth suggest the profound influence of digital media on theories, practices, research environments, and perhaps most importantly, significances attributed to and found within the humanities. Disciplinary traditions are in active interplay with the technologies even as the technologies are transforming the traditions, so it is more accurate to say that the tectonic shifts currently underway are technologically enabled and catalyzed rather than technologically driven, operating in recursive feedback loops rather than linear determinations. In broad view, the impact of these feedback loops is not confined to the humanities alone, reaching outward to redefine institutional categories, reform pedagogical practices, and reenvision the relation of higher education to local communities and global conversations.

If, as public opinion and declining enrollments might indicate, the Traditional Humanities are in trouble, entangled with this trouble is an opportunity. As Cathy N. Davidson (2008) argues, "We live in the information age. . . . I would insist that this is our age and that it is time we claimed it and engaged with it in serious, sustained, and systemic ways" (708). The Digital Humanities offer precisely this opportunity. Neither the Traditional nor the Digital Humanities can succeed as well alone as they can together. If the Traditional Humanities are at risk of becoming marginal to the main business of the contemporary academy and society, the Digital Humanities are at risk of becoming a trade practice held captive by the interests of corporate capitalism. Together, they offer an enriched, expanded repertoire of strategies, approaches, and assumptions that can more fully address the challenges of the information age than can either alone. By this I do not mean to imply that the way forward will be harmonious or easy. Nevertheless, the clash of assumptions between the Traditional and Digital Humanities presents an opportunity to rethink humanistic practices and values at a time when the humanities in general are coming under increasing economic and culture pressures. Engaging with the broad spectrum of issues raised by the Digital Humanities can help to ensure the continuing vitality and relevance of the humanities into the twenty-first century and beyond.

3

How We Read

Close, Hyper, Machine

The preceding chapter discussed the changes taking place within the humanities as a result of the increasing prevalence, use, and sophistication of digital media in academia. It was concerned mostly with activities central to the research mission of humanities faculty, with occasional comments on how teaching is affected as a result. This chapter looks at the other side of the coin, how digital media are affecting the practices in which our students are engaged, especially reading of digital versus print materials, with attention to the implications of these changes for pedagogy. Since the nature of cognition is centrally involved in these issues, this chapter also begins to develop a theory of embodied cognition encompassing conscious, unconscious, and nonconscious processes that will be crucial to the arguments of this and subsequent chapters.

The evidence is mounting: people in general, and young people in particular, are doing more screen reading of digital materials than ever before. Meanwhile, the reading of print books and of literary genres (novels, plays, and poems) has been declining over the last twenty years. Worse, reading skills (as measured by the ability to identify themes, draw inferences, etc.) have been declining in junior high, high school, college, and even graduate schools for the same period. Two flagship reports from the National Endowment for the Arts (NEA), *Reading at Risk* (2004), reporting the results of its own surveys, and *To Read or Not to Read* (2007), drawing together other large-scale surveys, show that over a wide range of data-gathering instruments the results are consistent: people read less print, and they read print less well. This leads NEA chairman Dana Gioia to suggest that the *correlation* between decreased literary reading and poorer reading ability is indeed a *causal* connection (NEA 2004). The NEA argues (and I of course agree) that literary reading is a good in itself, insofar as it opens the portals of a rich literary heritage (see Griswold, McDonnell, and Wright [2005] for the continued high cultural value placed on reading). When decreased print reading, already a cultural concern, is linked with reading problems, it carries a double whammy.

Fortunately, the news is not all bad. A newer NEA report, *Reading on the Rise* (2009), shows for the first time in more than two decades an uptick in novel reading (but not plays or poems), including among the digitally native young adult cohort (ages eighteen to twenty-four). The uptick may be a result of the "Big Read" initiative by the NEA and similar programs by other organizations; whatever the reason, it shows that print can still be an alluring medium. At the same time, reading scores among fourth and eighth graders remain flat, despite the "No Child Left Behind" initiative. The complexities of the national picture notwithstanding, it seems clear that a critical nexus occurs at the juncture of digital reading (exponentially increasing among all but the oldest cohort) and print reading (downward trending with a slight uptick recently). The crucial questions are these: how to convert the increased digital reading into increased reading ability, and how to make effective bridges between digital reading and the literacy traditionally associated with print.

Mark Bauerlein (a consultant on the *Reading at Risk* report) in the offensively titled *The Dumbest Generation: How the Digital Age Stupefies Young Americans and Jeopardizes Our Future* (2009) makes no apology for linking the decline of reading skills directly to a decrease in print reading, issuing a stinging indictment to teachers, professors, and other mentors who think digital reading might encourage skills of its own. Not only is there no transfer between digital reading and print reading skills in his view, but digital reading does not even lead to strong *digital* reading skills (2009:93–111). I found *The Dumbest Generation* intriguing and infuriating in equal measure. The book is valuable for its synthesis of a great deal of empirical evidence, going well beyond the 2008 NEA report in this regard; it is infuriating in its tendentious refusal to admit any salutary effects from digital reading. As Bauerlein moves from the solid longitudinal data on the decline in print reading to the digital realm, the evidence becomes scantier and the anecdotes more frequent, with examples obviously weighted toward showing the inanity of online chats, blogs, and Facebook entries. It would, of course, be equally possible to harvest examples showing the depth, profundity, and brilliance of online discourse, so Bauerlein's argument here fails to persuade. The two earlier NEA reports (2004, 2007) suffer from their own problems; their data do not clearly distinguish between print and digital reading, and they fail to measure how much digital reading is going on or its effects on reading abilities (Kirschenbaum 2007). Nevertheless, despite these limitations and distortions, few readers are likely to come away unconvinced that there is something like a national crisis in reading and that it is especially acute with teen and young adult readers.

At this point, scholars in literary studies should be jumping on their desks and waving their hands in the air, saying, "Hey! Look at us! We know how to read *really* well, and we know how to teach students to read. There's a national crisis in reading? We can help." Yet there is little evidence that the profession of literary studies has made a significant difference in the national picture, including on the college level, where reading abilities continue to decline even into graduate school. This is strange. The inability to address the crisis successfully no doubt has multiple causes, but one in particular strikes me as vitally important. While literary studies continues to teach close reading to students, it does less well in exploiting the trend toward the digital. Students read incessantly in digital media and write in it as well, but only infrequently are they encouraged to do so in literature classes or in environments that encourage the transfer of print reading abilities to digital and vice versa. The two tracks, print and digital, run side by side, but messages from either track do not leap across to the other.

Close Reading and Disciplinary Identity

To explore why this should be so and open possibilities for synergistic interactions, I begin by revisiting that sacred icon of literary studies, close reading. When literary studies expanded its purview in the 1970s and 1980s, it turned to reading many different kinds of "texts," from Donald Duck to fashion clothing, television programs to prison architecture (Scholes 1999). This expansion into diverse textual realms meant that literature was no longer the de facto center of the field. Faced with the loss of this traditional center, literary scholars found a replacement in close reading, the one thing virtually all literary scholars know how to do well and agree is important. Close reading then assumed a preeminent role as the essence of the disciplinary identity.

Jane Gallop undoubtedly speaks for many when she writes, "I would argue that the most valuable thing English ever had to offer was the very thing that made us a discipline, that transformed us from cultured gentlemen into a profession [i.e., close reading]. . . . Close reading—learned through practice with literary texts, learned in literature classes—is a widely applicable skill, of real value to students as well as to scholars in other disciplines" (2009:15). Barbara Johnson, in her well-known essay "Teaching Deconstructively" (1985), goes further: "This [close reading] is the only teaching that can properly be called literary; anything else is history of ideas, biography, psychology, ethics, or bad philosophy" (140). For Gallop, Johnson, and many others, close reading not only assures the professionalism of the profession but also makes literary studies an important asset to the culture. As such, close reading justifies the discipline's continued existence in the academy, as well as the monies spent to support literature faculty and departments. More broadly, close reading in this view constitutes the major part of the cultural capital that literary studies relies on to prove its worth to society.

Literary scholars generally think they know what is meant by close reading, but looked at more closely, it proves not so easy to define or exemplify. Jonathan Culler (2010), quoting Peter Middleton, observes that "close reading is our contemporary term for a heterogeneous and largely unorganized set of practices and assumptions" (20). John Guillory (2010a) is more specific when he historicizes close reading, arguing that "close reading is a modern academic practice with an inaugural moment, a period of development, and now perhaps a period of decline" (8). He locates its prologue in the work of I. A. Richards, noting that Richards contrasted close reading with the media explosion of his day, television. If that McLuhanesque view of media is the prologue, then digital technologies, Guillory suggests, may be launching the epilogue. Citing my work on hyper attention (more on that shortly), Guillory sets up a dichotomy between the close reading recognizable to most literary scholars—detailed and precise attention to rhetoric, style, language choice, and so forth through a word-by-word analysis of a text's linguistic techniques—to the digital world of fast reading and sporadic sampling. In this he anticipates the close versus digital reading flagrantly on display in Bauerlein's book.

Amid the heterogeneity of close reading techniques, perhaps the dominant one in recent years has been what Stephen Best and Sharon Marcus (2009) call "symptomatic reading." In a special issue of *Representations* titled "The Way We Read Now," Best and Marcus launch a frontal assault on symptomatic reading as it was inaugurated by Fredric Jameson's immensely influential *The Political Unconscious* (1981). For Jameson, with his motto "Always historicize," the text is an alibi for subtextual ideological formations. The heroic task of the critic is to wrench a text's ideology into the light, "massy and dripping" as Mary Crane puts it (2009:245), so that it can be unveiled and resisted (see Crane [2009] for a close analysis of Jameson's metaphors). The trace of symptomatic reading may be detected in Barbara Johnson. Listing textual features that merit special attention, she includes such constructions as "ambiguous words," "undecidable syntax," and "incompatibilities between what a text says and what it does" (1985:141–42). Most if not all of these foci are exactly the places where scholars doing symptomatic reading would look for evidence of a text's subsurface ideology.

After more than two decades of symptomatic reading, however, many literary scholars are not finding it a productive practice, perhaps because (like many deconstructive readings) its results have begun to seem formulaic, leading to predictable conclusions rather than compelling insights. In a paraphrase of Gilles Deleuze and Félix Guattari's famous remark, "We are tired of trees," the *Representations* special issue may be summarized as "We are tired of symptomatic reading." The issue's contributors are not the only ones who feel this way. In panel after panel at the conference sponsored by the National Humanities Center in spring 2010 entitled "The State and Stakes of Literary Studies," presenters expressed similar views and urged a variety of other reading modes, including "surface reading," in which the text is examined not for hidden clues but its overt messages; reading aimed at appreciation and articulation of the text's aesthetic value; and a variety of other reading strategies focusing on affect, pleasure, and cultural value.

Digital and Print Literacies

If one chapter of close reading is drawing to an end, what new possibilities are arising? Given the increase in digital reading, obvious sites for new kinds of reading techniques, pedagogical strategies, and initiatives are the interactions between digital and print literacies. Literary studies has been slow to address these possibilities, however, because it continues to view close reading of print texts as the field's essence. As long as this belief holds sway, digital reading will at best be seen as peripheral to our concerns, pushed to the margins as not "really" reading or at least not compelling or interesting reading. Young people, who vote with their feet in college, are marching in another direction—the digital direction. No doubt those who already read well will take classes based on close reading and benefit from them, but what about others whose print reading skills are not as highly developed? To reach them, we must start close to where they are, rather than where we imagine or hope they might be. As David Laurence (2008) observes, "Good teachers deliberately focus on what the reader can do, make sure that both teacher and student recognize and acknowledge it, and use it as a platform of success from which to build" (4).

This principle was codified by the Belarusian psychologist L. S. Vygotsky in the 1930s as the "zone of proximal development." In *Mind in Society: Development of Higher Psychological Processes* (1978), he defined this zone as "the distance between the actual developmental level as determined by independent problem solving and the level of potential development as determined through problem solving under adult guidance, or in collaboration with more capable peers" (86). The concept implies that if the distance is too great between what one wants someone else to learn and where instruction begins, the teaching will not be effective. Imagine, for example, trying to explain *Hamlet* to a three-year-old (an endless string of "Why?" would no doubt result, the all-purpose response of young children to the mysterious workings of the adult world). More recent work on "scaffolding" (Robertson, Fluck, and Webb n.d.) and on the "zone of reflective capacity" (Tinsley and Lebak 2009) extends the idea and amplifies it with specific learning strategies. These authors agree that for learning to occur, the distance between instruction and available skills must be capable of being bridged, either through direct instruction or, as Vygotsky notes, through working with "more capable" peers. Bauerlein instances many responses from young people as they encounter difficult print texts to the effect the works are "boring" or not worth the trouble. How can we convey to such students the deep engagement we feel with challenging literary texts? I argue that we cannot do this effectively if our teaching does not take place in the zone of proximal development, that is, if we are focused exclusively on print close reading. Before opinion solidifies behind new versions of close reading, I want to argue for a disciplinary shift to a broader sense of reading strategies and their interrelation.

James Sosnoski (1999) presciently introduced the concept of hyper reading, which he defined as "reader-directed, screen-based, computer-assisted reading" (167). Examples include search queries (as in a Google search), filtering by keywords, skimming, hyperlinking, "pecking" (pulling out a few items from a longer text), and fragmenting (Sosnoski 1999:163–72). Updating his model, we may add juxtaposing, as when several open windows allow one to read across several texts, and scanning, as when one reads rapidly through a blog to identify items of interest. There is considerable evidence that hyper reading differs significantly from typical print reading, and moreover that hyper reading stimulates different brain functions than print reading.

For example, Jakob Nielson and his consulting team, which advises companies and others on effective web design, does usability research by asking test subjects to deliver running verbal commentaries as they encounter web pages. Their reactions are recorded by a (human) tester; at the same time, eye-tracking equipment records their eye movements. The research shows that web pages are typically read in an F pattern (Nielson 2006). A person reads the first two or three lines across the page, but as the eye travels down the screen, the scanned length gets smaller, and by the time the bottom of the page is reached, the eye is traveling in a vertical line aligned with the left margin. (Therefore the worst location for important information on a web page is on the bottom right corner.) In Bauerlein's view, this research confirms that digital reading is sloppy in the extreme; he would no doubt appreciate Woody Allen's quip, "I took a speed reading course and was able to read *War and Peace* in twenty minutes. It involves Russia" (qtd. in Dehaene 2009:18). Nevertheless, other research not cited by Bauerlein indicates that this and similar strategies work well to identify pages of interest and to distinguish them from pages with little or no relevance to the topic at hand (Sillence et al. 2007).

As a strategic response to an information-intensive environment, hyper reading is not without precedent. John Guillory, in "How Scholars Read" (2008), notes that "the fact of quantity is an intractable empirical given that must be managed by a determined method if analysis or interpretation is to be undertaken" (13). He is not talking here about digital reading but about archival research that requires a scholar to move through a great deal of material quickly to find the relevant texts or passages. He identifies two techniques in particular, scanning (looking for a particular keyword, image, or other textual feature) and skimming (trying to get the gist quickly). He also mentions the book wheel, a physical device invented in the Renaissance to cope with the information explosion when the number of books increased exponentially with the advent of print. Resembling a five-foot-high Ferris wheel, the book wheel held several books on different shelves and could be spun around to make different texts accessible, in a predigital print version of hyper reading.

In contemporary digital environments, the information explosion of the web has again made an exponentially greater number of texts available, dwarfing the previous amount of print materials by several orders of magnitude. In digital environments, hyper reading has become a necessity. It enables a reader quickly to construct landscapes of associated research fields and subfields; it shows ranges of possibilities; it identifies texts and passages most relevant to a given query; and it easily juxtaposes many different texts and passages. Google searches and keyword filters are now as much part of the scholar's toolkit as hyper reading itself. Yet hyper reading may not sit easily alongside close reading. Recent studies indicate that hyper reading not only requires different reading strategies than close reading but also may be involved with changes in brain architecture that makes close reading more difficult to achieve.

Much of this evidence is summarized by Nicholas Carr in *The Shallows: What the Internet Is Doing to Our Brains* (2010). More judicious than Bauerlein, he readily admits that web reading has enormously increased the scope of information available, from global politics to scholarly debates. He worries, however, that hyper reading leads to changes in brain function that make sustained concentration more difficult, leaving us in a constant state of distraction in which no problem can be explored for very long before our need for continuous stimulation kicks in and we check e-mail, scan blogs, message someone, or check our RSS feeds. The situation is reminiscent of Kurt Vonnegut's satirical short story "Harrison Bergeron" ([1961] 1998), in which the pursuit of equality has led to a society that imposes handicaps on anyone with exceptional talents. The handsome, intelligent eponymous protagonist must among other handicaps wear eyeglasses that give him headaches; other brainiacs have radio transmitters implanted in their ears, which emit shrieking sounds two or three times every minute, interrupting their thoughts and preventing sustained concentration. The final satirical punch comes in framing the story from the perspective of Bergeron's parents, Hazel and George, who see their son on TV when he proclaims his antihandicap manifesto (with fatal results for him), but, hampered by their own handicaps, they cannot concentrate enough to remember it.

The story's publication in 1961 should give us a clue that a media-induced state of distraction is not a new phenomenon. Walter Benjamin, in "The Work of Art in the Age of Mechanical Reproduction" (1968b), wrote about the ability of mass entertainment forms such as cinema (as opposed to the contemplative view of a single work of art) to make distracted viewing a habit. Even though distraction, as Jonathan Crary (2001) has shown, has been a social concern since the late 1800s, there are some new features of web reading that make it a powerful practice for rewiring the brain (see Greenfield [2009]). Among these are hyperlinks that draw attention away from the linear flow of an article, very short forms such as tweets that encourage distracted forms of reading, small habitual actions such as clicking and navigating that increase the cognitive load, and, most pervasively, the enormous amount of material to be read, leading to the desire to skim everything because there is far too much material to pay close attention to anything for very long.

Reading on the Web

What evidence indicates that these web-specific effects are making distraction a contemporary cultural condition? Several studies have shown that, contrary to the claims of early hypertext enthusiasts such as George Landow, hyperlinks tend to degrade comprehension rather than enhance it. The following studies, cited by Carr in *The Shallows* (2010), demonstrate the trend. Erping Zhu (1999), coordinator of instructional development at the Center for Research on Learning and Teaching at the University of Michigan, had test subjects read the same online passage but varied the number of links. As the number of links increased, comprehension declined, as measured by writing a summary and completing a multiple-choice test. Similar results were found by two Canadian scholars, David S. Miall and Teresa Dobson (2001), who asked seventy people to read Elizabeth Bowen's short story "The Demon Lover." One group read it in a linear version and a second group with links. The first group outperformed the second on comprehension and grasp of the story's plot; it also reported liking the story more than the second group. We may object that a print story would of course be best understood in a print-like linear mode; other evidence, however, indicates that a similar pattern obtains for digital-born material. D. S. Niederhauser and others (2000) had test subjects read two online articles, one arguing that "knowledge is objective" and the other that "knowledge is relative." Each article had links allowing readers to click between them. The researchers found that those who used the links, far from gaining a richer sense of the merits and limitations of the two positions, understood them less well than readers who chose to read the two in linear fashion. Comparable evidence was found in a review of thirty-eight experiments on hypertext reading by Diana DeStefano and Jo-Anne LeFevre (2007), psychologists with the Centre for Applied Cognitive Research at Canada's Carleton University. Carr summarizes their results, explaining that, in general, the evidence did not support the claim that hypertext led to "an enriched experience of the text" (qtd. in Carr 2010:129). One of DeStefano and LeFevre's conclusions was that "increased demands of decision-making and visual processing in hypertext impaired reading performance," especially in relation to "traditional print presentation" (qtd. in Carr 2010:129).

Why should hypertext, and web reading in general, lead to poorer comprehension? The answer, Carr believes, lies in the relation of working memory (i.e., the contents of consciousness) to long-term memory. Material is held in working memory for only a few minutes, and the capacity of working memory is severely limited. For a simple example, I think of the cell phone directory function that allows me to get phone numbers, which are given orally (there is an option to have a text message sent of the number, but for this the company charges an additional fee, and being of a frugal disposition, I don't go for that option). I find that if I repeat the numbers out loud several times so they occupy working memory to the exclusion of other things, I can retain them long enough to punch the number. For retention of more complex matters, the contents of working memory must be transferred to long-term memory, preferably with repetitions to facilitate the integration of the new material with existing knowledge schemas. The small distractions involved with hypertext and web reading—clicking on links, navigating a page, scrolling down or up, and so on—increase the cognitive load on working memory and thereby reduce the amount of new material it can hold. With linear reading, by contrast, the cognitive load is at a minimum, precisely because eye movements are more routine and fewer decisions need to be made about how to read the material and in what order. Hence the transfer to long-term memory happens more efficiently, especially when readers reread passages and pause to reflect on them as they go along.

Supplementing this research are other studies showing that small habitual actions, repeated over and over, are extraordinarily effective in creating new neural pathways. Carr recounts the story told by Norman Doidge in *The Brain That Changes Itself* of an accident victim who had a stroke that damaged his brain's right side, rendering his left hand and leg crippled (Carr 2010:30–31). He entered an experimental therapy program that had him performing routine tasks with his left arm and leg over and over, such as washing a window, tracing alphabet letters, and so forth. "The repeated actions," Carr reports, "were a means of coaxing his neurons and synapses to form new circuits that would take over the functions once carried out by the circuits in the damaged area in his brain" (2010:30). Eventually, the patient was able to regain most of the functionality of his unresponsive limbs. We may remember in the film *The Karate Kid* (1984) when Daniel LaRusso (Ralph Macchio) is made to do the same repetitive tasks over and over again by his karate teacher, Mr. Miyagi (Pat Morita). In contemporary neurological terms, Mr. Miyagi is retraining the young man's neural circuits so he can master the essentials of karate movements.

These results are consistent with a large body of research on the impact of (print) reading on brain function. In a study cited by the French neurophysiologist Stanislas Dehaene (2009), a world-renowned expert in this area, researchers sought out siblings from poor Portuguese families who had followed the traditional custom of having an elder sister stay home and watch the infant children, while her younger sister went to school. Raised in the same family, the sisters could be assumed to have grown up in very similar environments; the pairing thus served as a way to control other variables. Using as test subjects six pairs of illiterate/literate sisters, researchers found that literacy had strengthened the ability to understand the phonemic structure of language. Functional magnetic resonance imaging (fMRI) scans showed pronounced differences in the anterior insula, adjacent to Broca's area (a part of the brain associated with language use). "The literate brain," Dehaene summarizes, "obviously engages many more left hemispheric resources that the illiterate brain, even when we only *listen* to speech. . . . The macroscopic finding implies a massive increase in the exchange of information across the two hemispheres" (2009:209).

Equally intriguing is Dehaene's "neural recycling" hypothesis, which suggests that reading repurposes existing brain circuits that evolved independently of reading (because literacy is a mere eye blink in our evolutionary history, it did not play a role in shaping the genetics of our Pleistocene brains but rather affects us epigenetically through environmental factors). Crucial in this regard is an area he calls the brain's "letterbox," located in the left occipito-temporal region at the back of the brain. This area, fMRI data show, is responsible for letter and phonemic recognition, transmitting its results to other distant areas through fiber bundles. He further argues that brain architecture imposes significant constraints on the physical shapes that will be easily legible to us. He draws on research demonstrating that 115 of the world's diverse writing systems (alphabetic and ideographic) use visual symbols consisting mostly of three strokes (plus or minus one). Moreover, the geometry of these strokes mirrors in their distribution the geometry of shapes in the natural environment. The idea, then, is that our writing systems evolved in the context of our ability to recognize natural shapes; scribal experimentation used this correspondence to craft writing systems that would most effectively repurpose existing neural circuitry. Dehaene thus envisions "a massive selection process: over time, scribes developed increasingly efficient notations that fitted the organization of our brains. In brief, our cortex did not specifically evolve for writing. Rather, writing evolved to fit the cortex" (2009:171).

Current evidence suggests that we are now in a new phase of the dance between epigenetic changes in brain function and the evolution of new reading and writing modalities on the web. Think, for example, of the F pattern of web reading that Nielson's research revealed. Canny web designers use this information to craft web pages, and reading such pages further intensifies this mode of reading. How quickly neural circuits may be repurposed by digital reading is suggested by Gary Small's experiments at the University of California, Los Angeles, on the effects of web reading on brain functionality. Small and his colleagues were looking for digitally naïve subjects; they recruited three volunteers in their fifties and sixties who had never performed Google searches (Small and Vorgan 2008:15–17). This group was first tested with fMRI brain scans while wearing goggles onto which were projected web pages. Their scans differed significantly from another group of comparable age and background who were web savvy. Then the naïve group was asked to search the Internet for an hour a day for five days. When retested, their brain scans showed measurable differences in some brain areas, which the experimenters attributed to new neural pathways catalyzed by web searching. Citing this study among others, Carr concludes that "knowing what we know today, if you were to set out to invent a medium that would rewire our mental circuits as quickly and thoroughly as possible, you would probably end up designing something that looks and works a lot like the Internet" (2010:116).

How valid is this conclusion? Although Carr's book is replete with many different kinds of studies, we should be cautious about taking his conclusions at face value. For example, in the fMRI study done by Small and his colleagues, many factors might skew the results. I don't know if you have had a brain scan, but I have. As Small mentions, brain scans require that you be shoved into a tube just big enough to accommodate your supine body but not big enough for you to turn over. When the scan begins, supercooled powerful electromagnets generate a strong magnetic field, which, combined with a radio frequency emitter, allows minute changes in blood oxygen levels in the brain to be detected and measured. When the radio frequency emitter begins pulsing, it sounds as though a jackhammer is ripping up pavement next to your ear. These are hardly typical conditions for web reading. In addition, there is considerable evidence that fMRI scans, valuable as they are, are also subject to a number of interpretive errors and erroneous conclusions (Sanders 2009). Neural activity is not measured directly by fMRI scans (as a microelectrode might, for example). Rather, the most widely used kind of fMRI, BOLD (blood oxygen level dependent), measures tiny changes in oxygenated blood as a correlate for brain activity. BOLD research assumes that hardworking neurons require increased flows of oxygen-rich blood and that protons in hemoglobin molecules carrying oxygen respond differently to magnetic fields than protons in oxygen-depleted blood. These differences are tabulated and then statistically transformed into colored images, with different colors showing high levels of oxygen-rich compared to depleted-oxygen blood.

The chain of assumptions that led Small, for example, to conclude that brain function changed as a result of Google searches can go wrong in several different ways (see Sanders [2009] for a summary of these criticisms). First, researchers assume that the *correlation* between activity in a given brain area is *caused* by a particular stimulus; however, most areas of the brain respond similarly to several different kinds of stimuli, so another stimulus could be activating the change rather than the targeted one. Second, fMRI data sets typically have a lot of noise, and if the experiment is not repeated, the observed phenomenon may be a chimera rather than a genuine result (in Small's case, the experiment was repeated later with eighteen additional volunteers). Because the data sets are large and unwieldy, researchers may resort to using sophisticated statistical software packages they do not entirely understand. In addition, the choice of colors used to visualize the statistical data is arbitrary, and different color contrasts may cause the images to be interpreted differently. Finally, researchers may be using a circular methodology in which the hypothesis affects how the data are seen (an effect called nonindependence). When checkers went back through fMRI research that had been published in the premier journals *Nature, Science, Nature Neuroscience, Neuron*, and the *Journal of Neuroscience*, they found interpretive errors resulting from nonindependence in 42 percent of the papers (cited in Sanders 2009:16).

Relying on summaries of research in books such as Carr's creates additional hazards. I mentioned earlier a review of hypertext experiments (DeStefano and LeFevre 2007) cited by Carr, which he uses to buttress his claim that hypertext reading is not as good as linear reading. Consulting the review itself reveals that Carr has tilted the evidence to support his view. DeStefano and LeFevre state, for example, that "there may be cases in which enrichment or complexity of the hypertext experience is more desirable than maximizing comprehension and ease of navigation," remarking that this may be especially true for students who already read well. They argue not for abandoning hypertext but rather for "good hypertext design" that takes cognitive load into account "to ensure hypermedia provide *at least as good* a learning environment as more traditional text" (2007:1636; emphasis added). Having read through most of Carr's primary sources, I can testify that he is generally conscientious in reporting research results; nevertheless, the example illustrates the unsurprising fact that reading someone else's synthesis does not give as detailed or precise a picture as reading the primary sources themselves.

The Importance of Anecdotal Evidence

Faced with these complexities, what is a humanist to do? Obviously, few scholars in the humanities have the time—or the expertise—to backtrack through cited studies and evaluate them for correctness and replicability. In my view, these studies may be suggestive indicators but should be subject to the same kind of careful scrutiny we train our students to use with web research (reliability of sources, consensus among many different researchers, etc.). Perhaps our most valuable yardstick for evaluating these results, however, is our own experience. We know how we react to intensive web reading, and we know through repeated interactions with our students how they are likely to read, write, and think as they grapple with print and web materials. As teachers (and parents), we make daily observations that either confirm or disconfirm what we read in the scientific literature. The scientific research is valuable and should not be ignored, but our experiences are also valuable and can tell us a great deal about the advantages and disadvantages of hyper reading compared with close reading, as well as the long-term effects of engaging in either or both of these reading strategies.

Anecdotal evidence hooked me on this topic five years ago, when I was a Phi Beta Kappa Scholar for a year and in that capacity visited many different types of colleges and universities. Everywhere I went, I heard teachers reporting similar stories: "I can't get my students to read long novels anymore, so I've taken to assigning short stories"; "My students won't read long books, so now I assign chapters and excerpts." I hypothesized then that a shift in cognitive modes is taking place, from the deep attention characteristic of humanistic inquiry to the hyper attention characteristic of someone scanning web pages (Hayles 2007a). I further argued that the shift in cognitive modes is more pronounced the younger the age cohort. Drawing from anecdotal evidence as well as such surveys as the Kaiser Foundation's "Generation M" report (Roberts, Foehr, and Rideout 2005), I suggested that the shift toward hyper attention is now noticeable in college students. Since then, the trend has become even more apparent, and the flood of surveys, books, and articles on the topic of distraction is now so pervasive as to be, well, distracting.[1]

For me, the topic is much more than the latest research fad, because it hits me where I live: the college classroom. As a literary scholar, I deeply believe in the importance of writing and reading, so any large-scale change in how young people read and write is bound to capture my attention. In my work on hyper attention (published just when the topic was beginning to appear on the national radar), I argued that deep and hyper attention each have distinctive advantages. Deep attention is essential for coping with complex phenomena such as mathematical theorems, challenging literary works, and complex musical compositions; hyper attention is useful for its flexibility in switching between different information streams, its quick grasp of the gist of material, and its ability to move rapidly among and between different kinds of texts.[2] As contemporary environments become more information-intensive, it is no surprise that hyper attention (and its associated reading strategy, hyper reading) is growing and that deep attention (and its correlated reading strategy, close reading) is diminishing, particularly among young adults and teens. The problem, as I see it, lies not in hyper attention and hyper reading as such but rather in the challenges the situation presents for parents and educators to ensure that deep attention and close reading continue to be vibrant components of our reading cultures and interact synergistically with the kind of web and hyper reading in which our young people are increasingly immersed.[3]

Yet hyper reading and close reading are not the whole story. I earlier referred to Sosnoski's definition of hyper reading as "computer-assisted." More precisely, it is computer-assisted human reading. The formulation alerts us to a third component of contemporary reading practices: human-assisted computer reading, that is, computer algorithms used to analyze patterns in large textual corpora where size makes human reading of the entirety impossible. We saw in chapter 2 that machine reading ranges from algorithms for word frequency counts to more sophisticated programs that find and compare phrases, identify topic clusters, and are capable of learning. In 2006, Oren Etzioni, Michele Banko, and Michael J. Cafarella (2006) at the University of Washington argued that the time was ripe to launch an initiative on machine reading, given contemporary advances in natural language processing, machine learning, and probabilistic reasoning. They define machine reading as "the automatic, unsupervised understanding of text" (1). Elaborating, Etzioni writes, "By 'understanding text' I mean the formation of a coherent set of beliefs based on a textual corpus and a background theory. Because the text and the background theory may be inconsistent, it is natural to express the resultant beliefs, and the reasoning process in probabilistic terms" (Etzioni et al. 2006:1). Necessary to achieve this goal are programs that can draw inferences from text: "A key problem is that many of the beliefs of interest are only *implied* by the text in combination with a background theory" (1; emphasis in original). In 2007, the American Association of Artificial Intelligence organized the first symposium on machine reading, including papers on textual inferences, semantic integration, ontology learning, and developing and answering questions (Etzioni 2007).

A similar project at Carnegie Mellon University is the Never-Ending Language Learning (NELL) program, directed by Tom M. Mitchell and his group of graduate students in computer science. In their technical report, they identify a reading task, "extract information from a web text to further populate a growing knowledge base of structured facts and knowledge," and a learning task, "learn to read better each day than the day before, as evidenced by its ability to go back to yesterday's text sources and extract more information more accurately" (Carlson et al. 2010:1). The computer program consists of four modules, including a "Rule Learner" and a "Coupled Pattern Learner" that extracts instances of categories and relations from text "in the wild" (that is, the immense body of texts on the web unrestricted by imposed constraints or canonical limits). From these modules, the program constructs "candidate facts"; based on a high degree of confidence from one module or lower degrees of confidence from several modules, it then elevates some of the candidates to "beliefs." The program operates 24/7 and moreover is iterative, constantly scouring the web for text and constructing relations of the kind "X is an instance of Y which is a Z." The researchers initially seeded the program's knowledge base with 123 categories and 55 relations; after 67 days, the knowledge base contained 242,453 new facts with an estimated precision of 74 percent. There is a risk in this procedure, for the program tests new candidate facts for consistency with facts already in the knowledge base. If an incorrect fact makes it into the knowledge base, then, it tends to encourage the approval of more incorrect facts. To correct this tendency, human readers check the "Rule Learner" and other program modules for ten to fifteen minutes each day to correct errors the program does not correct for itself. As a result, NELL is not an unsupervised system but a "semi-supervised" program. A recent visit to the website (October 10, 2010, http://rtw.ml.cmu.edu/rtw/) revealed these "recently learned facts": "a golden-bellied euphonia is a bird," and "vastus medialis is a muscle." The program's errors are as revealing as its correct inferences. The same visit revealed these gems: "property offences is a kind of military event," "englishes is an ethnic group," and my favorite, "english is the language of the country Japan." Thinking about what information is on the web regarding English and Japan, for example, makes the latter inference understandable if also incorrect.

Although machine reading as a technical field is still in its infancy, the potential of the field for the construction of knowledge bases from unstructured text is clear. Moreover, in unsupervised or lightly supervised programs, machine reading has the huge advantage of never sleeping, never being distracted by other tasks. In a field like literary studies, the efficacy, scope, and importance of machine reading are widely misunderstood. Even such a perceptive critic as Jonathan Culler falls back on caricature when, in writing about close reading, he suggests, "It may be especially important to reflect on the varieties of close reading and even to propose explicit models, in an age where electronic resources make it possible to do literary research without reading at all: find all the instances of the words *beg* and *beggar* in novels by two different authors and write up your conclusions" (2010:241). His emphasis on close reading is admirable (also typical of literary studies), but his implication that drawing conclusions from machine analysis ("write up your conclusions") is a mechanical exercise devoid of creativity, insight, or literary value is far off the mark. Even John Guillory, a brilliant theorist and close reader, while acknowledging that machine reading is a useful "prosthesis for the cognitive skill of scanning," concludes that "the gap in cognitive level between the keyword search and interpretation is for the present immeasurable" (2008:13). There are two misapprehensions here: that keyword searches exhaust the repertoire of machine reading and that the gap between analysis and interpretation yawns so wide as to form an unbridgeable chasm rather than a dynamic interaction. As we saw in chapter 2, the line between (human) interpretation and (machine) pattern recognition is a very porous boundary, with each interacting with the other. As demonstrated through many examples there, hypotheses about meaning help shape the design of computer algorithms (the "background theory" referred to above), and the results of algorithmic analyses refine, extend, and occasionally challenge intuitions about meaning that form the starting point for algorithmic design. Putting human reading in a leak-proof container and isolating machine reading in another makes it difficult to see these interactions and understand their complex synergies. Given these considerations, saying computers cannot read is from my point of view merely species chauvinism.

In view of these misconceptions, explicit recapitulation of the value of machine reading is useful. Although machine reading may be used with a single text and reveal interesting patterns (see for example the coda in chapter 8), its more customary use is in analyzing corpora too vast to be read by a single person, as Gregory Crane notes in his remarks cited in chapter 2. The NELL program, as another example, takes the endless ocean of information on the web as its data source. More typical in literary studies are machine algorithms used to read large corpora of literary texts, which although very large are still much smaller and often much more structured than the free text on the web. The problem tackled by researchers such as Franco Moretti, although complex, are nevertheless not as difficult as the kind of challenge taken on by NELL, because Moretti's data are frequently tagged with metadata and presented in more or less standard formats. Reading text in the wild, as NELL does, understandably has a lower probability of accurate inferences than reading more structured or limited textual canons.

Before going further, I want to add a note on nomenclature. Moretti (2007) uses the term "distant reading," an obvious counterpoise to close reading. Careful reading of his work reveals that this construction lumps together human and machine reading; both count as "distant" if the scale is large. I think it is useful to distinguish between human and machine reading because the two situations (one involving a human assisted by machines, the other involving computer algorithms assisted by humans) have different functionalities, limitations, and possibilities. Hyper reading may not be useful for large corpora, and machine algorithms have limited interpretive capabilities.

If we look carefully at Moretti's methodology, we see how firmly it refutes the misunderstandings referred to above. His algorithmic analysis is usually employed to pose questions. Why are the lifetimes of many different genres limited to about thirty years (Moretti 2007)? Why do British novels in the mid-eighteenth century use many words in a title and then, within a few decades, change so that titles are usually no more than three or four words long (Moretti 2009)? Allen Beye Riddell has pointed out that Moretti is far more interested in asking provocative questions than in checking his hypotheses; Moretti's attention, Riddell argues, is directed toward introducing a new paradigm rather than providing statistically valid answers. Nevertheless, Moretti's explanations can on occasion be startlingly insightful, as in his analysis of what happens to free indirect discourse when the novel moves from Britain to British colonies (Moretti 2007). When the explanations fail to persuade (as Moretti candidly confesses is sometimes the case even for him), the patterns nevertheless stand revealed as entry points for interpretations advanced by other scholars who find them interesting.

I now turn to explore the interrelations between the components of an expanded repertoire of reading strategies that includes close, hyper, and machine reading. The overlaps between them are as revealing as the differences. Close and hyper reading operate synergistically when hyper reading is used to identity passages or to home in on a few texts of interest, whereupon close reading takes over. As Guillory observes, skimming and scanning here alternate with in-depth reading and interpretation (2008). Hyper reading overlaps with machine reading in identifying patterns. This might be done in the context of a Google keyword search, for example, when one notices that most of the work on a given topic has been done by X, or it might be when machine analysis confirms a pattern already detected by hyper (or close) reading. Indeed, skimming, scanning, and pattern identification are likely to occur in all three reading strategies; their prevalence in one or another is a matter of scale and emphasis rather than clear-cut boundary.

Since patterns have now entered the discussion, we may wonder what a pattern is. This is not a trivial question, largely because of the various ways in which patterns become manifest. Patterns in large data sets may be so subtle that only sophisticated statistical analysis can reveal them; complex patterns may nevertheless be apprehended quickly and easily when columns of numbers are translated into visual forms, as with fMRI scans. Verbal patterns may be discerned through the close reading of a single textual passage or grasped through hyper reading of an entire text or many texts. An anecdote may be useful in clarifying the nature of pattern. I once took a pottery class, and the instructor asked each participant to make several objects that would constitute a series. The series might, for example, consist of vases with the same shapes but different sizes, or it might be vases of the same size in which the shapes underwent a consistent set of deformations. The example shows that differences are as important as similarities, for they keep a pattern from being merely a series of identical items. I therefore propose the following definition: a pattern consists of regularities that appear through a series of related differences and similarities.

Related to the idea of pattern is the question of meaning. Since entire books have been written on the subject, I will not attempt to define meaning but merely observe that wherever and however it occurs, meaning is sensitively dependent on context. The same sentence, uttered in two different contexts, may mean something entirely different in one than in the other. Close reading typically occurs in a monolocal context (that is, with a single text). Here the context is quite rich, including the entire text and other texts connected with it through networks of allusions, citations, and iterative quotations. Hyper reading, by contrast, typically occurs in a multilocal context. Because many textual fragments are juxtaposed, context is truncated, often consisting of a single phrase or sentence, as in a Google search. In machine reading, the context may be limited to a few words or eliminated altogether, as in a word-frequency list. Relatively context poor, machine reading is enriched by context-rich close reading when close reading provides guidance for the construction of algorithms; Margaret Cohen points to this synergy when she observes that for computer programs to be designed, "the patterns still need to be observed [by close reading]" (2009:59). On the other hand, machine reading may reveal patterns overlooked in close reading, a point we saw in chapter 2 with Willard McCarty's work on personification in Ovid's *Metamorphoses* (2005:3–72). The more the emphasis falls on pattern (as in machine reading), the more likely it is that context must be supplied from outside (by a human interpreter) to connect pattern with meaning; the more the emphasis falls on meaning (as in close reading), the more pattern assumes a subordinate role. In general, the different distributions among pattern, meaning, and context provide ways to think about interrelations among close, hyper, and machine reading.

The larger point is that close, hyper, and machine reading each have distinctive advantages and limitations; nevertheless, they also overlap and can be made to interact synergistically with one another. Maryanne Wolfe reaches a similar conclusion when, at the end of *Proust and the Squid* (2007), she writes, "We must teach our children to be 'bitextual' or 'multitextual,' able to read and analyze texts flexibly in different ways, with more deliberate instruction at every stage of development on the inferential, demanding aspects of any text. Teaching children to uncover the invisible world that resides in written words needs to be both explicit and part of a dialogue between learner and teacher, if we are to promote the processes that lead to fully formed expert reading in our citizenry" (226). I agree wholeheartedly with the goal; the question is how, precisely, to accomplish it.

Synergies between Close, Hyper, and Machine Reading

Starting from a traditional humanistic basis in literature, Alan Liu in the English Department at the University of California, Santa Barbara, has been teaching undergraduate and graduate courses that he calls "Literature+," which adopt as a pedagogical method the interdisciplinarity facilitated by digital media (A. Liu 2008c). He asks students "to choose a literary work and treat it according to one or more of the research paradigms prevalent in other fields of study," including visualization, storyboarding, simulation, and game design. Starting with close reading, he encourages students to compare it with methodologies in other fields, including the sciences and engineering. He also has constructed a "Toy Chest" on his website that includes links to software packages enabling students with little or no programming experience to create different modes of representation of literary texts, including tools for text analysis, visualization, mapping, and social-network diagramming. The approach is threefold: it offers students traditional literary training; it expands their sense of how they can use digital media to analyze literary texts; and it encourages them to connect literary methodologies with those of other fields they may be entering. It offers close reading not as an unquestioned good but as one methodology among several, with distinctive capabilities and limitations. Moreover, because decisions about how to encode and analyze texts using software programs require precise thinking about priorities, goals, and methodologies, it clarifies the assumptions that undergird close reading by translating them into algorithmic analysis.

An example of how the "Literature+" approach works in practice is the project entitled "*Romeo and Juliet*: A Facebook Tragedy" (Skura, Nierle, and Gin 2008). Three students working collaboratively adapted Shakespeare's play to the Facebook model, creating maps of social networks using the

"Friend Wheel" (naturally, the Montagues are all "friends" to each other, and so are the Capulets), filling out profiles for the characters (Romeo is interpreted as a depressive personality who has an obsessive attachment to his love object and corresponding preferences for music, films, and other cultural artifacts that express this sensibility), and having a fight break out on the message board forum using a "Group" called "The Streets of Verona." The "Wall" feature was used to incorporate dialogue in which characters speak directly to one another, and the "Photos" section allowed one character to comment on the attributes of another. The masque at which Romeo and Juliet meet became an "Event," to which Capulet invited friends in his "Friend Wheel." From a pedagogical point of view, the students were encouraged to use software with which they were familiar in unfamiliar ways, thus increasing their awareness of its implications. The exercise also required them to make interpretive judgments about which features of the play were most essential (since not everything could be included) and to be precise about interactions between relationships, events, and characters. Linking traditional literary reading skills with digital encoding and analysis, the Literature+ approach strengthens the ability to understand complex literature at the same time it encourages students to think reflectively on digital capabilities. Here digital and print literacies mutually reinforce and extend one another.

Lev Manovich's "Cultural Analytics" (2007) is a series of projects that start from the premise that algorithmic analyses of large data sets (up to several terabytes in size), originally developed for work in the sciences and social sciences, should be applied to cultural objects, including the analysis of real-time data flows. In many academic institutions, high-end computational facilities have programs that invite faculty members and graduate students in the arts and humanities to use them. For example, at the University of California, San Diego, where Manovich teaches, the Supercomputer Center sponsored a summer workshop in 2006 titled Cyberinfrastructure for the Humanities, Arts and Social Sciences. At Duke University, where I teach, the Renaissance Computing Institute offers accounts to faculty members and students in the arts and humanities that allow them to do computationally intense analysis. In my experience, researchers at these kinds of facilities are delighted when humanists come to them with projects. Because their mission is to encourage widespread use across and among campuses and to foster collaborations among academic, government, corporate, and community stakeholders, they see humanistic inquiry and artistic creation as missing parts of the picture that enrich the mix. This opens the door to analysis of large cultural data sets such as visual images, media content, and geospatial mapping combined with various historical and cultural overlays.

An example is Manovich's analysis of *Time* magazine covers from 1923 to 1989 (Manovich 2007). As Manovich observes, ideal sites for cultural analytics are large data sets that are well structured and include metadata about date, publication venue, and so forth. The visualization tools that he uses allow the *Time* covers to be analyzed according to subject (for example, portraits versus other types of covers), color gradients, black-and-white gradients, amount of white space, and so forth. One feature is particularly useful for building bridges between close reading and machine analysis: the visualization tool allows the user both to see large-scale patterns and to zoom in to see a particular cover in detail, thus enabling analyses across multiple scale levels. Other examples include Manovich's analysis of one million manga pages using the Modrian software, sorted according to gray-scale values; another project analyzes scene lengths and gray-scale values in classic black-and-white films. While large-scale data analyses are not new, their applications in the humanities and arts are still in their infancy, making cultural analytics a frontier of knowledge construction.

Of course, not everyone has access to computation-intensive facilities, including most parents and teachers at smaller colleges and universities. A small-scale example that anyone could implement will be helpful. In teaching an honors writing class, I juxtaposed Mary Shelley's *Frankenstein* with Shelley Jackson's *Patchwork Girl*, an electronic hypertext fiction written in proprietary Storyspace software. Since these were honors students, many of them had already read *Frankenstein* and were, moreover, practiced in close reading and literary analysis. When it came to digital reading, however, they were accustomed to the scanning and fast skimming typical of hyper reading; they therefore expected that it might take them, oh, half an hour to go through Jackson's text. They were shocked when I told them a reasonable time to spend with Jackson's text was about the time it would take them to read *Frankenstein*, say, ten hours or so. I divided them into teams and assigned a section of Jackson's text to each team, telling them that I wanted them to discover *all* the lexias (i.e., blocks of digital text) in their section and warning them that the Storyspace software allows certain lexias to be hidden until others are read. Finally, I asked them to diagram interrelations between lexias, drawing on all three views that the Storyspace software enables.

As a consequence, the students were not only required to read closely but also to analyze the narrative strategies Jackson uses to construct her text. Jackson focuses some of her textual sections on a narrator modeled on the female creature depicted in *Frankenstein*, when Victor, at the male creature's request, begins to assemble a female body as a companion to his first creation (Hayles 2001). As Victor works, he begins to think about the two creatures mating and creating a race of such creatures. Stricken with sexual nausea, he tears up the female body while the male creature watches, howling, from the window; throws the pieces into a basket; and rows out to sea, where he dumps them. In her text Jackson reassembles and reanimates the female creature, playing with the idea of fragmentation as an inescapable condition not only for her narrator but for all humans. The idea is reinforced by the visual form of the narrative, which (in the Storyspace map view) is visualized as a series of titled text blocks connected by webs of lines. Juxtaposing this text with *Frankenstein* encouraged discussions about narrative framing, transitions, strategies, and characterization. By the end, the students, who already admired *Frankenstein* and were enthralled by Mary Shelley's narrative, were able to see that electronic literature might be comparably complex and would also repay close attention to its strategies, structure, form, rhetoric, and themes. Here already-existing print literacies were enlisted to promote and extend digital literacy.

These examples merely scratch the surface of what can be done to create productive interactions between close, hyper, and machine reading. Close and hyper reading are already part of a literary scholar's toolkit (although hyper reading may not be recognized or valued as such). Many good programs are now available for machine reading, such as Wordle, which creates word clouds to display word frequency analysis; RapidMiner, which enables collocation analyses; and the advanced version of the Hermetic Word Frequency counter, which has the ability to count words in multiple files and to count phrases as well as words. Other text analysis tools are available through the Text Analysis Portal for Research (TAPoR) text-analysis portal (http://portal.tapor.ca/portal/coplets/myprojects/taporTools/). Most of these programs are not difficult to use and provide the basis for wide-ranging experimentation by students and teachers alike. As Manovich says about cultural analytics and Moretti proclaims about distant reading, machine analysis opens the door to new kinds of discoveries that were not possible before and that can surprise and intrigue scholars accustomed to the delights of close reading.

What transformed disciplinary coherence might literary studies embrace? The Comparative Media Studies approach provides a framework within which an expanded repertoire of reading strategies may be pursued. Here is a suggestion for how these might be described: "Literary studies teaches literacies across a range of media forms, including print and digital, and focuses on interpretation and analysis of patterns, meaning, and context through close, hyper, and machine reading practices." Reading has always been constituted through complex and diverse practices. Now it is time to rethink what reading is and how it works in the rich mixtures of words and images, sounds and animations, graphics and letters that constitute the environments of twenty-first century literacies.

SECOND INTERLUDE

The Complexities of Contemporary Technogenesis

Contemporary technogenesis, like evolution in general, is not about progress. That is, it offers no guarantees that the dynamic transformations taking place between humans and technics are moving in a positive direction. Rather, contemporary technogenesis is about adaptation, the fit between organisms and their environments, recognizing that both sides of the engagement (humans and technologies) are undergoing coordinated transformations. Already this situation is more complex than a more simplistic Darwinian scenario that sees the environment as static while change happens mostly within the organism. The complexities do not end there, however, for the instruments by which one might attempt to measure these changes are themselves part of the technical environment and so are also involved in dynamic transformations. The situation is akin to a relativistic scenario of a spaceship traveling at near light speed: the clocks on board by which one might measure time dilation are themselves subject to the very phenomenon in question, so accurate measurement of dilation effects by this means is impossible. Needed are approaches broad enough to capture the scope of the changes underway, flexible enough to adapt to changes in criteria implied by technogenetic transformations, and subtle enough to distinguish between positive and negative outcomes when the very means by which such judgments are made may themselves be in question.

This, of course, is a tall order. The strategy this book follows is to start with specific sites such as the Digital Humanities and reading, the subjects of chapters 2 and 3, where the emphasis falls mostly on how digital media are changing the human practices of writing and reading, teaching and learning. Chapter 4 offers a theoretical perspective on how change happens within technics, concluding with analyses that braid together technological and human adaptations. Chapter 5 then moves outward to a more sweeping analysis of large-scale cultural changes. Since it is always easier to see the import of changes that have happened in the past than those taking place in the present, this large-scale analysis focuses on the historical technologies of telegraphy and telegraph code books.

By the conclusion of chapter 5, the reader will have been presented with a substantial body of evidence indicating that contemporary technogenesis through digital media is indeed a dominant trend in the United States (and by extension, in other developed countries), that it has historical roots going back to the nineteenth century, that how we think about temporality is deeply involved with these changes, that academic practices in the humanities as well as more general processes within the culture are undergoing dynamic transformations in coordination with digital media, and that evaluating these changes in political, social, and ethical terms requires careful consideration of the complexities involved.

At the same time, questions of how to use these understandings to make constructive changes in the world are at least as complex as the understandings themselves. Catherine Malabou offers a provocative solution in *What Should We Do with Our Brain?* (2008). We should, she proposes, use neural plasticity to become conscious of our own possibilities for self-fashioning, including explosive subversion of the flexibility that is the hallmark of contemporary global capitalism. Flexibility implies subservience to the dominant New World Order, whereas plasticity implies the capacity not only to make but to unmake, to explosively resist as well as to adapt. Careful analysis of her argument indicates that she locates this agential freedom in the gap or rupture between the neuronal level of representation (the "proto-self," as Antonio Damasio calls it) and the mental level of conscious thought (the narrated or autobiographical self). It is precisely because this connection remains somewhat mysterious, she argues, that cognitive science (and indeed any of the objective sciences including neurophysiology, neurology, and related disciplines) must go beyond merely objective description and take an ideologically informed stance such as she advocates. The problem with this analysis, however, is that the postulated gap between the neuronal and conscious self is, by definition, not available to conscious representation. How can conscious awareness that such a gap exists (assuming it does) bring about the agential freedom she envisions, and what practical actions are possible that would take advantage of this freedom?

Although Malabou recognizes that plasticity involves an active dynamic between environmental forces and neuronal changes, she mentions media only once in her discussion, and that is to refer to "an afflicting economic, political, and *mediatric* culture that celebrates only the triumph of flexibility, blessing obedient individuals who have no greater merit than that of knowing how to bow their heads with a smile" (2008:79; emphasis added). This condescending view of media (presumably including digital media) forecloses an important resource for contemporary self-fashioning, for using plasticity both to subvert and redirect the dominant order. It is precisely because contemporary technogenesis posits a strong connection between ongoing dynamic adaptation of technics and humans that multiple points of intervention open up. These include making new media (recall Timothy Lenoir's comment, "We make media. That's what we do"); adapting present media to subversive ends (a strategy Rita Raley discusses in *Tactical Media* [2009]); using digital media to reenvision academic practices, environments, and strategies (the subject of chapter 2); and crafting reflexive representations of media self-fashionings in electronic and print literatures that call attention to their own status as media, in the process raising our awareness of both the possibilities and dangers of such self-fashioning (discussed in chapters 4, 7, and 8). This list is obviously not exhaustive, but I hope that it illustrates enough of the possibilities to indicate how digital media can be used to intervene constructively in our present situation and how these interventions are enabled by the technogenetic spiral, even when they aim to subvert or resist it.

4

Tech-TOC

Complex Temporalities and Contemporary Technogenesis

Chapters 2 and 3 took as their entry points into contemporary technogenesis the changes in human attitudes, assumptions, and cognitive modes associated with digital media. This chapter advances a theoretical framework in which technical objects are also seen in evolutionary terms as repositories of change and, more profoundly, as agents and manifestations of complex temporalities. To explore the coordinated epigenetic dynamic between humans and technics, this chapter focuses on the kinds of temporal scales involved when humans and digital media interact. To arrive at that point, it is first necessary to think about how temporality manifests itself within technical objects.

At least since Henri Bergson's concept of duration, a strong distinction has been drawn between temporality as process (according to Bergson, unextended, heterogeneous time at once multiplicitous and unified, graspable only through intuition and human experience) and temporality as measured (homogenous, spatialized, objective, and "scientific" time) (Bergson [1913] 2005). Its contributions to the history of philosophy notwithstanding, the distinction has a serious disadvantage: although objects, like living beings, exist within duration, there remains a qualitative distinction between the human capacity to grasp duration and the relations of objects to it. Indeed, there can be no account of how duration is experienced by objects, for lacking intuition, they may manifest duration but not experience it. What would it mean to talk about an object's experience of time, and what implications would flow from this view of objecthood?

Exploring these questions requires an approach that can recognize the complexity of temporality with regard to technical objects. This would open the door to a series of provocative questions. How is the time of objects constructed as complex temporality? What human cognitive processes participate in this construction? How have the complex temporalities of objects and humans coconstituted one another through epigenetic evolutionary processes? Along what time scales do interactions occur between humans and technical objects, specifically networked and programmable machines? What are the implications of concatenating processual and measured time together in the context of digital technologies? What artistic and literary strategies explore this concatenation, and how does their mediation through networked and programmable machines affect the concatenation?

Nearly half a century ago, the French "mechanologist" Gilbert Simondon ([1958] 2001) proposed an evolutionary explanation for the regularities underlying transformations of what he called "technical beings." Simondon's analyses remain useful primarily because of his focus on general principles combined with detailed expositions of how they apply to specific technologies. Working through Simondon's concepts as well as their elaboration by such contemporary theorists as Adrian MacKenzie, Mark Hansen, Bernard Stiegler, and Bruno Latour, I will discuss the view that technical objects embody complex temporalities enfolding past into present, present into future. An essential component of this approach is a shift from seeing technical objects as static entities to conceptualizing them as temporary coalescences in fields of conflicting and cooperating forces. While hints of Bergsonian duration linger here, the mechanologist perspective starts from a consideration of the nature of technical objects rather than from the nature of human experience.

Working from previous discussions of human-centered change, this chapter combines with them the object-centered view, thus putting into place the other half necessary to launch the full technogenetic spiral. In this dynamic, attention plays a special role, for it focuses on the detachment and reintegration of technical elements that Simondon argues is the essential mechanism for technological change. Its specialness notwithstanding, attention is nevertheless a limited cognitive faculty whose boundaries are difficult to see because it is at the center of consciousness, whereas unconscious and nonconscious faculties remain partially occluded, despite being as (or more) important in interactions with technological environments. Embodied human cognition as a whole (including attentive focus, unconscious perceptions, and nonconscious cognitions) provides the basis for dynamic interactions with the tools it helps to bring into being.

While a mechanologist perspective, combined with human-centered technogenesis, provides an explanatory framework within which complex temporalities can be seen to inhabit both living and technical beings, the pervasive human experience of clock time cannot be so easily dismissed. To explore the ways in which duration and spatialized temporality create fields of contention seminal to human cultures, I turn to an analysis of *TOC: A New Media Novel* (2009), a multimodal electronic novel by Steve Tomasula (with design by Stephen Farrell). Marked by aesthetic ruptures, discontinuous narratives, and diverse modalities including video, textual fragments, animated graphics, and sound, *TOC* creates a rich assemblage in which the conflicts between measured and experienced time are related to the invention, development, and domination of the "Influencing Engine," a metaphoric allusion to computational technologies. Composed on computers and played on them, *TOC* explores its conditions of possibility in ways that perform as well as demonstrate the interpenetration of living and technical beings, processes in which complex temporalities play central roles.

Technics and Complex Temporalities

Technics, in Simondon's view, is the study of how technical objects emerge, solidify, disassemble, and evolve. In his exposition, technical objects comprise three different categories: technical elements, for example, the head of a stone ax; technical individuals, for example, the compound tool formed by the head, bindings, and shaft of a stone ax; and technical ensembles, in the case of an ax, the flint piece used to knap the stone head, the fabrication of the bindings from animal material, and the tools used to shape the wood shaft, as well as the toolmaker who crafts this compound tool. The technical ensemble gestures toward the larger social/technical processes through which fabrication comes about; for example, the toolmaker is himself embedded in a society in which the knowledge of how to make stone axes is preserved, transmitted, and developed further. Obviously, it is difficult to establish clear-cut boundaries between a technical ensemble and the society that creates it, leading to Bruno Latour's observation that a tool is "the extension of social skills to non-humans" (1999:211). Whereas the technical ensemble points toward complex social structures, the technical element shows how technological change can come about by being detached from its original ensemble and embedded in another, as when the stone ax head becomes the inspiration for a stone arrowhead. The ability of technical elements to "travel" is, in Simondon's view, a principal factor in technological change.

The motive force for technological change is the increasing tendency toward what Simondon calls "concretization," innovations that resolve conflicting requirements within the milieu in which a technical individual operates. Slightly modifying his example of an adze, I illustrate conflicting requirements in the case of the metal head of an ax. The ax blade must be hard enough so that when it is driven into a piece of cord wood, for example, it will cut the wood without breaking. The hole through which the ax handle goes, however, has to be flexible enough so that it can withstand the force transmitted from the blade without cracking. The conflicting requirements of flexibility and rigidity are negotiated through the process of tempering, in which the blade is hammered so that the crystal structure is changed to give it the necessary rigidity, while the thicker part of the head retains its original properties.

Typically, when a technical individual is in early stages of development, conflicting requirements are handled in an ad hoc way, for example in the separate system used in a water-cooled internal combustion engine that involves a water pump, radiator, tubing, etc. When problems are solved in this fashion, the technical individual is said to be abstract, in the sense that it instantiates separate bodies of knowledge without integrating them into a unitary operation. In an air-cooled engine, by contrast, the cooling becomes an intrinsic part of the engine's operation; if the engine operates, the cooling necessarily takes place. As an example of increasing concretization, Simondon instances exterior ribs that serve both to cool a cylinder head and simultaneously provide structural support so that the head will remain rigid during repeated explosions and impacts. Concretization, then, is at the opposite end of the spectrum from abstraction, for it integrates conflicting requirements into multipurpose solutions that enfold them together into intrinsic and necessary circular causalities. Simondon correlates the amount of concretization of a technical individual with its technicity; the greater the degree of concretization, the greater the technicity.

The conflicting requirements of a technical individual that is substantially abstract constitute, in Simondon's view, a potential for innovation or, in Deleuzian terms, a repository of virtuality that invites transformation. Technical individuals represent, in Adrian Mackenzie's phrase, metastabilities, that is, provisional solutions of problems whose underlying dynamics push the technical object toward further evolution (2002:17ff.). Detaching a technical element from one individual—for example, a transistor from a computer—and embedding it in a new technical ensemble or individual—for example, importing it into a radio, cell phone, or microwave—requires solutions that move the new milieu toward greater concretization. These solutions sometimes feed back into the original milieu of the computer to change its functioning as well. In this view, technical objects are always on the move toward new configurations, new milieu, and new kinds of technical ensembles. Temporality is something that not only happens within them but also is carried by them in a constant dance of temporary stabilizations amid continuing innovations.

This dynamic has implications for the "folding of time," a phenomenon Bruno Latour (1994) identifies as crucial to understanding technological change. Technical ensembles, as we have seen, create technical individuals; they are also called into existence by technical individuals. The automobile, for example, called a national network of paved roads into existence; it in turn was called into existence by the concentration of manufacture in factories, along with legal regulations that required factories to be located in separate zones from housing. In this way, the future is already preadopted in the present (future roads in present cars), while the present carries along with it the marks of the past, for example in the metal ax head that carries in its edge the imprint of the technical ensemble that tempered it (in older eras, this would include a blacksmith, forge, hammer, anvil, bucket of water, etc.). On a smaller scale, the past is enfolded into the present through skeuomorphs, details that were previously functional but have lost their functionality in a new technical ensemble. The "stitching" (actually an impressed pattern) on my Honda Accord vinyl dashboard covering, for example, recalls the older leather coverings that were indeed stitched. So pervasive are skeuomorphs that once one starts looking for them, they appear to be everywhere: why, if not for the human need to carry the past into the present? These enfoldings—past nestling inside present, present carrying the embryo of the future—constitute the complex temporalities that inhabit technics.

When devices are created that make these enfoldings explicit, for example in the audio and video recording devices that Bernard Stiegler (1998:173–79; 2009:78–79) discusses, the biological capacity for memory (which can be seen as an evolutionary adaptation to carry the past into the present) is exteriorized, creating the possibility, through technics, for a person to experience through complex temporality something that never was experienced as a firsthand event, a possibility Stiegler calls tertiary retention (or tertiary memory). This example, which Stiegler develops at length and to which he gives theoretical priority, should not cause us to lose sight of the more general proposition: that all technics imply, instantiate, and evolve through complex temporalities.

The Coevolution of Humans and Tools

Simondon distinguishes between a technical object and a (mere) tool by noting that technical objects are always embedded within larger networks of technical ensembles, including geographic, social, technological, political, and economic forces. For my part, I find it difficult to imagine a tool, however humble, that does not have this characteristic. Accordingly, I depart from Simondon by considering tools as part of technics, and indeed an especially important category because of their capacity for catalyzing exponential change. Anthropologists usefully define a tool as an artifact used to make other artifacts. The definition makes clear that not everything qualifies as a tool: houses, cars, and clothing are not normally used to make other artifacts, for example. One might suppose that the first tools were necessarily simple in construction, because by definition, there were no other tools to aid in their fabrication. In the new millennium, by contrast, the technological infrastructure of tools that make other artifacts is highly developed, including laser cutters, 3-D printers, and the like, as well as an array of garden-variety tools that populate the garages of many US households, including power saws, nail hammers, cordless drills, etc. Arguably, the most flexible, pervasive, and useful tools in the contemporary era are networked and programmable machines. Not only are they incorporated into an enormous range of other tools, from computerized sewing machines to construction cranes, but they also produce informational artifacts as diverse as databases and Flash poems.

The constructive role of tools in human evolution involves cognitive as well as muscular and skeletal changes. Stanley Ambrose (2001), for example, has linked the fabrication of compound tools (tools with more than one part, such as a stone ax) to the rapid increase in Broca's area in the brain and the consequent expansion and development of language. According to his argument, compound tools involve specific sequences for their construction, a type of reasoning associated with Broca's area, which also is instrumental in language use, including the sequencing associated with syntax and grammar. Tool fabrication in this view resulted in cognitive changes that facilitated the capacity for language, which in turn further catalyzed the development of more (and more sophisticated) compound tools.

To move from such genetic changes to contemporary technogenesis, we may begin by bringing to the fore a cognitive faculty undertheorized within the discourse of technics: the constructive role of attention in fabricating tools and creating technical ensembles. To develop this idea, I distinguish between materiality and physicality. The physical attributes of any object are essentially infinite (indeed, this inexhaustible repository is responsible for the virtuality of technical objects). As I discuss in *My Mother Was a Computer: Digital Subjects and Literary Texts* (2005), a computer can be analyzed through any number of attributes, from the rare metals in the screen to the polymers in the case to the copper wires in the power cord, and so on ad infinitum. Consequently, physical attributes are necessary but not sufficient to account for technical innovation. What counts is rather the object's *materiality*. Materiality comes into existence, I argue, when attention fuses with physicality to identify and isolate some particular attribute (or attributes) of interest.

Materiality is unlike physicality in being an emergent property. It cannot be specified in advance, as though it existed ontologically as a discrete entity. Requiring acts of human attentive focus on physical properties, materiality is a human-technical hybrid. Matthew Kirschenbaum's definitions of materiality in *Mechanisms: New Media and the Forensic Imagination* (2008) may be taken as examples illustrating this point. Kirschenbaum distinguishes between forensic and formal materiality. Forensic materiality, as the name implies, consists in minute examinations of physical evidence to determine the traces of information in a physical substrate, as when, for example, one takes a computer disk to a nanotech laboratory to "see" the bit patterns (with a nonoptical microscope). Here attention is focused on determining one set of attributes, and it is this fusion that allows Kirschenbaum to detect (or more properly, construct) the materiality of the bit pattern. Kirschenbaum's other category, formal materiality, can be understood through the example of the logical structures in a software program. Formal materiality, like forensic materiality, must be embodied in a physical substrate to exist (whether as a flowchart scribbled on paper, a hierarchy of coding instructions written into binary within the machine, or the firing of neurons in the brain), but the focus now is on the form or structure rather than physical attributes. In both cases, forensic and formal, the emergence of materiality is inextricably bound up with the acts of attention. Attention also participates in the identification, isolation, and modification of technical elements that play a central role in the evolution of technical objects.

At the same time, other embodied cognitive faculties also participate in this process. Andy Pickering's (1995) description of the "mangle of practice" illustrates. Drawing on my own experience with throwing pots, I typically would begin with a conscious idea: the shape to be crafted, the size, texture, and glaze, etc. Wedging the clay gives other cognitions a chance to work as I absorb through my hands information about the clay's graininess, moisture content, chemical composition, etc., which may perhaps cause me to modify my original idea. Even more dynamic is working the clay on the wheel, a complex interaction between what I envision and what the clay has a mind to do. A successful pot emerges when these interactions become a fluid dance, with the clay and my hands coming to rest at the same moment. In this process, embodied cognitions of many kinds participate, including unconscious and nonconscious ones running throughout my body and, through the rhythmic kicking of my foot, extending into the wheel. That these capacities should legitimately be considered a part of cognition is eloquently argued by Antonio Damasio (2005), supplemented by the work of Oliver Sacks (1998) on patients who have suffered neurological damage and so have, for example, lost the capacity for memory, proprioception, or emotion. As these researchers demonstrate, losing these capacities results in cognitive deficits so profound that they prevent the patients from living anything like a normal life. Conversely, normal life requires embodied cognition that exceeds the boundaries of consciousness and indeed of the body itself.

Embedded cognition is closely related to extended cognition, with different emphasis and orientation that nevertheless are significant. An embedded cognitive approach, typified by the work of anthropologist Edwin Hutchins (1996), emphasizes the environment as crucial scaffolding and support for human cognition. Analyzing what happened when a navy vessel lost steam power and then electric power as it entered San Diego harbor and consequently lost navigational capability, Hutchins observed the emergence of self-organizing cognitive systems in which people made use of the devices to hand (a ruler, paper, and protractor, for example) to make navigational calculations. No one planned in advance how these systems would operate, because there was no time to do so; rather, people under extreme pressure simply did what made sense in that environment, adapting to each other's decisions and to the environment in flexible yet coordinated fashion. In this instance, and many others that Hutchins analyzes, people make use of objects in the environment, including their spatial placements, colors, shapes, and other attributes, to support and extend memory, combine ideas into novel syntheses, and in general enable more sophisticated thoughts than would otherwise be possible.

Andy Clark illustrates this potential with a story (qtd. from Gleick, 1993: 409) about Richard Feynman, the Nobel Prize–winning physicist, meeting with the historian Charles Weiner to discuss a batch of Feynman's original notes. Weiner remarks that the papers are "a record of [Feynman's] day-to-day work," but Feynman disagrees.

> "I actually did the work on the paper," he said.
>
> "Well," Weiner said, "the work was done in your head, but the record of it is still here."
>
> "No, It's not a *record*, not really. It's *working*. You have to work on paper and this is the paper. Okay?" (Clark 2008:xxv).

Feynman makes clear that he did not have the ideas in advance and wrote them down. Rather, the process of writing down was an integral part of his thinking, and the paper and pencil were as much a part of his cognitive system as the neurons firing in his brain. Working from such instances, Clark develops the model of extended cognition (which he calls EXTENDED), contrasting it with a model that imagines cognition happens only in the brain (which he calls BRAINBOUND). The differences between EXTENDED and BRAINBOUND are clear, with the neurological, experimental, and anecdotal evidence overwhelmingly favoring the former over the latter.

More subtle are the differences between the embedded and extended models. Whereas the embedded approach emphasizes human cognition at the center of self-organizing systems that support it, the extended model tends to place the emphasis on the cognitive system as a whole and its enrollment of human cognition as a part of it. Notice, for example, the place of human cognition in this passage from Clark: "We should consider the possibility of a vast parallel coalition of more or less influential forces, whose largely self-organizing unfolding makes each of us the thinking beings we are" (2008:131–32). Here human agency is downplayed, and the agencies of "influential forces" seem primary. In other passages, human agency comes more to the fore, as in the following: "Goals are achieved by making the most of reliable sources of relevant order in the bodily or worldly environment of the controller" (2008:5–6). On the whole, however, it is safe to say that human agency becomes less a "controller" in Clark's model of extended cognition compared to embedded cognition and more of a player among many "influential forces" that form flexible, self-organizing systems of which it is a part. It is not surprising that there should be ambiguities in Clark's analyses, for the underlying issues involve the very complex dynamics between deeply layered technological built environments and human agency in both its conscious and unconscious manifestations. Recent work across a range of fields interested in this relation—neuroscience, psychology, cognitive science, and others—indicates that the unconscious plays a much larger role than had previously been thought in determining goals, setting priorities, and other activities normally associated with consciousness. The "new unconscious," as it is called, responds in flexible and sophisticated ways to the environment while remaining inaccessible to consciousness, a conclusion supported by a wealth of experimental and empirical evidence.

For example, John A. Bargh (2005) reviews research in "behavior-concept priming." In one study, university students were presented with what they were told was a language test; without their knowledge, one of the word lists was seeded with synonyms for rudeness, the other with synonyms for politeness. After the test, participants exited via a hallway in which there was a staged situation, to which they could react either rudely or politely. The preponderance of those who saw the rude words acted rudely, while those who saw the polite words acted politely. A similar effect has been found in activating stereotypes. Bargh writes, "Subtly activating (priming) the professor stereotype in a prior context causes people to score higher on a knowledge quiz, and priming the elderly stereotype makes college students not only walk more slowly but have poorer incidental memory as well" (2005:39). Such research demonstrates "the existence of sophisticated nonconscious monitoring and control systems that can guide behavior over extended periods of time in a changing environment, in pursuit of desired goals" (43). In brain functioning, this implies that "conscious intention and behavioral (motor) systems are fundamentally dissociated in the brain. In other words, the evidence shows that much if not most of the workings of the motor systems that guide action are opaque to conscious access" (43).

To explain how nonconscious actions can pursue complex goals over an extended period of time, Bargh instances research indicating that working memory (about which we heard in chapter 3 as that portion of memory immediately accessible to consciousness, often called the brain's "scratch pad") is not a single unitary structure but rather has multiple components. He summarizes, "The finding that within working memory, representations of one's intentions (accessible to conscious awareness) are stored in a different location and structure from the representations used to guide action (not accessible) is of paramount importance to an understanding of the mechanisms underlying priming effects in social psychology" (47), providing "the neural basis for nonconscious goal pursuit and other forms of unintended behavior" (48). In this view, the brain remembers in part through the action circuits within working memory, but these memories remain beyond the reach of conscious awareness.

Given the complexity of actions that people can carry out without conscious awareness, the question arises of what is the role of consciousness, in particular, what evolutionary driver vaulted primate brains into self-awareness. Bargh suggests that "metacognitive awareness," being aware of what is happening in the environment as well as one's thoughts, allows the coordination of different mental states and activities to get them all working together. "Metacognitive consciousness is the workplace where one can assemble and combine the various components of complex perceptual-motor skills" (53). Quoting Merlin Donald (2001), he emphasizes the great advantage this gives humans: " 'whereas most other species depend on their built-in demons to do their mental work for them, *we can build our own demons*' " (qtd. in Bargh 2005:53; emphasis in original).

Nevertheless, many people resist the notion that nonconscious and unconscious actions may be powerful sources of cognition, no doubt because it brings human agency into question. This is a mistake, argue Ap Dijksterhuis, Henk Aarts, and Pamela K. Smith (2005). In a startling reversal of Descartes, they propose that thought itself is mostly unconscious. "Thinking about the article we want to write is an unconscious affair," they claim. "We read and talk, but only to acquire the necessary materials for our unconscious mechanisms to chew on. We are consciously aware of some of the products of thought that sometimes intrude into consciousness . . . but not of the thinking—the chewing—itself" (82). They illustrate their claim that "we should be happy that thought is unconscious" with statistics about human processing capacity. The senses can handle about 11 million bits per second, with about 10 million bits per second coming from the visual system. Consciousness, by contrast, can handle dramatically fewer bits per second. Silent reading processes take place at about 45 bits per second; reading aloud slows the rate to 30 bits per second. Multiplication proceeds at 12 bits per second.

Thus they estimate that "our total capacity is 200,000 times as high as the capacity of consciousness. In other words, consciousness can only deal with a very small percentage of all incoming information. All the rest is processed without awareness. Let's be grateful that unconscious mechanisms help out whenever there is a real job to be done, such as thinking" (82).

The Technological Unconscious

With unconscious and nonconscious motor processes assuming expanded roles in these views, we can now trace the cycles of continuous reciprocal causality that connect embodied cognition with the technological infrastructure (i.e., the built environment). Nigel Thrift (2005) argues that contemporary technical infrastructures, especially networked and programmable machines, are catalyzing a shift in the technological unconscious, that is, the actions, expectations, and anticipations that have become so habitual they are "automatized," sinking below conscious awareness while still being integrated into bodily routines carried on without conscious awareness. Chief among them are changed constructions of space (and therefore changed experiences of temporality) emerging with "track-and-trace" devices. With technologies such as bar codes, SIM cards in mobile phones, and radio frequency identification (RFID) tags, human and nonhuman actants become subject to hypercoordination and microcoordination (see Hayles 2009 for a discussion of RFID). Both time and space are divided into smaller and smaller intervals and coordinated with locations and mobile addresses of products and people, resulting in "a new kind of phenomenality of position and juxtaposition" (Thrift 2005:186). The result, Thrift suggests, is "a background sense of highly complex systems simulating life because, in a self-fulfilling prophecy . . . highly complex systems (of communication, logistics, and so on) *do* structure life and increasingly do so adaptively" (186). Consequently, mobility and universally coordinated time subtly shift what is seen as human. "The new phenomenality is beginning to structure what is human by disclosing 'embodied' capacities of communication, memory, and collaborative reach . . . that privilege a roving engaged interaction as typical of 'human' cognition and feed that conception back into the informational devices and environments that increasingly surround us" (186). "Human" in this construction is far from Lear's "unaccommodated man," "a poor, bare, forked animal"; rather, "human" in developed countries now means (for those who have access) cognitive capacities that extend into the environment, tap into virtually limitless memory storage, navigate effortlessly by

GPS, and communicate in seconds with anyone anywhere in the world (who also has access).

Although these changes are accelerating at unprecedented speeds, they have antecedents well before the twentieth century. Wolfgang Shivelbush (1987), for example, discusses them in the context of the railway journey, when passengers encountered landscapes moving faster than they had ever appeared to move before. He suggests that the common practice of reading a book on a railway journey was a strategy to cope with this disorienting change, for it allowed the passenger to focus on a stable nearby object that remained relatively stationary, thus reducing the anxiety of watching what was happening out the window. Over time, passengers came to regard the rapidly moving scenery as commonplace, an indication that the mechanisms of attention had become habituated to faster-moving stimuli.

An analogous change happened in films between about 1970 and the present. Film directors accept as common wisdom that the time it takes for an audience to absorb and process an image has decreased dramatically as jump cuts, flashing images, and increased paces of image projection have conditioned audiences to recognize and respond to images faster than was previously the case. My colleague Rita Raley tells of showing the film *The Parallax View* and having her students find it unintentionally funny, because the supposedly subliminal images that the film flashes occur so slowly (to them) that it seems incredible anyone could ever have considered them as occurring at the threshold of consciousness. Steven Johnson, in *Everything Bad Is Good for You* (2006), notes a phenomenon similar to faster image processing when he analyzes the intertwining plot lines of popular films and television shows to demonstrate that narrative development in these popular genres has become much more complicated in the last four decades, with shorter sequences and many more plot lines. These developments hint at a dynamic interplay between the kinds of environmental stimuli created in information-intensive environments and the adaptive potential of cognitive faculties in concert with them. Moreover, the changes in the environment and cognition follow similar trajectories, toward faster paces, increased complications, and accelerating interplays between selective attention, the unconscious, and the technological infrastructure.

It may be helpful at this point to recapitulate how this dynamic works. The "new unconscious," also called the "adaptive unconscious" by Timothy Wilson (2002), a phrase that seems to me more appropriate, creates the background that participates in guiding and directing selective attention. Because the adaptive unconscious interacts flexibly and dynamically with the environment (i.e., through the technological unconscious), there is a mediated relationship between attention and the environment much broader and more inclusive than focused attention itself allows. A change in the environment results in a change in the technological unconscious and consequently in the background provided by the adaptive unconscious, and that in turn creates the possibility for a change in the content of attention. The interplay goes deeper than this, however, for the *mechanisms of attention* themselves mutate in response to environmental conditions. Whenever dramatic and deep changes occur in the environment, attention begins to operate in new ways.

Andy Clark (2008) gestures toward such changes when, from the field of environmental change, he focuses on particular kinds of interactions that he calls "epistemic actions," which are "actions designed to change the input to an agent's information-processing system. They are ways an agent has of modifying the environment to provide crucial bits of information just when they are needed most" (38). I alter Clark's formulation slightly so that epistemic actions, as I use the term, are understood to modify *both* the environment and cognitive-embodied processes that adapt to make use of those changes. Among the epistemic changes in the last fifty years in developed countries such as the United States are dramatic increases in the use and pacing of media, including the web, television, and films; networked and programmable machines that extend into the environment, including PDAs, cell phones, GPS devices, and other mobile technologies; and the interconnection, data scraping, and accessibility of databases through a wide variety of increasingly powerful desktop machines as well as such ubiquitous technologies such as RFID tags, often coupled autonomously with sensors and actuators. In short, the variety, pervasiveness, and intensity of information streams have brought about major changes in built environments in the United States and comparably developed societies in the last half century. We would expect, then, that conscious mechanisms of attention and those undergirding the adaptive unconscious have changed as well. Catalyzing these changes have been new kinds of embodied experiences (virtual reality and mixed reality, for example), new kinds of cognitive scaffolding (computer keyboards, multitouch affordances in Windows 7), and new kinds of extended cognitive systems (cell phones, video games, multiplayer role games, and persistent reality environments such as Second Life, etc.). As Mark B. N. Hansen argues in *Bodies in Code* (2006a), the "new technical environments afford nothing less than the opportunity to *suspend* habitual causal patterns and, subsequently, to *forge* new patterns through the medium of embodiment—that is, by tapping into the flexibility (or potentiality) that characterizes humans as fundamentally embodied creatures" (29; emphasis in original).

As discussed in chapter 3, these environment changes have catalyzed changes in the mechanisms of selective attention from deep to hyper attention. Elsewhere I have argued that we are in the midst of a generational shift in cognitive modes (Hayles 2007a). If we look for the causes of this shift, media seem to play a major role. Empirical studies such as the "Generation M" report in 2005 by the Kaiser Family Foundation (Roberts, Foehr, and Rideout 2005) indicate that young people (ages eight to eighteen in their survey) spend, on average, an astonishing six hours per days consuming some form of media, and often multiple forms at once (surfing the web while listening to an iPod, for example, or texting their friends). Moreover, media consumption by young people in the home has shifted from the living room to their bedrooms, a move that facilitates consuming multiple forms of media at once. Going along with the shift is a general increase in information intensity, with more and more information available with less and less effort.

It is far too simplistic to say that hyper attention represents a cognitive deficit or a decline in cognitive ability among young people (see Bauerlein [2009] for an egregious example of this). On the contrary, hyper attention can be seen as a positive adaptation that makes young people better suited to live in the information-intensive environments that are becoming ever more pervasive. That being said, I think deep attention is a precious social achievement that took centuries, even millennia, to cultivate, facilitated by the spread of libraries, better K–12 schools, more access to colleges and universities, and so forth. Indeed, certain complex tasks can be accomplished only with deep attention: it is a heritage we cannot afford to lose. I will not discuss further here why I think the shift toward hyper attention constitutes a crisis in pedagogy for our colleges and universities, or what strategies might be effective in guiding young people toward deep attention while still recognizing the strengths of hyper attention (discussed to some extent in chapter 3). Instead, I want now to explore the implications of the shift for contemporary technogenesis.

Neurologists have known for some time that an infant, at birth, has more synapses (connections between neurons) than she or he will ever have again in life (see Bear, Connors, and Paradiso 2007). Through a process known as synaptogenesis, synapses are pruned in response to environmental stimuli, with those that are used strengthening and the neural clusters with which they are associated spreading, while the synapses that are not used wither and die. Far from cause for alarm, synaptogenesis can be seen as a marvelous evolutionary adaptation, for it enables every human brain to be reengineered from birth on to fit into its environment. Although greatest during the first two years of life, this process (and neural plasticity in general) continues throughout childhood and into adulthood. The clear implication is that children who grow up in information-intensive environments will literally have brains wired differently than children who grow up in other kinds of cultures and situations. The shift toward hyper attention is an indication of the direction in which contemporary neural plasticity is moving in developed countries. It is not surprising that it should be particularly noticeable in young people, becoming more pronounced the younger the child, down at least to ages three or four, when the executive function that determines how attention works is still in formation.

Synaptogenesis is one mechanism driving the change in selective attention, along with others discussed in chapter 3. The contemporary changes in mechanisms of attention are conveyed not through changes in the DNA but rather through epigenetic (i.e., environmental) changes. The relation between epigenetic and genetic changes has been a rich field of research in recent decades, resulting in a much more nuanced (and accurate) picture of biological adaptations than was previously understood, when the central dogma had all adaptations occurring through genetic processes. As mentioned in chapter 1, Mark James Baldwin (1896), a pioneer in epigenetic evolutionary theory, proposed in the late nineteenth century an important modification to Darwinian natural selection. He suggested that a feedback loop operates between genetic and epigenetic change, when populations of individuals that have undergone genetic mutation modify their environments so as to favor that adaptation. There are many examples in different species of this kind of effect (Deacon 1998; Clark 2004).

In the contemporary period, when epigenetic changes are moving young people toward hyper attention, a modified Baldwin effect may be noted (modified because the feedback loop here does not run between genetic mutation and environmental modification, as Baldwin proposed, but rather between epigenetic changes and further modification of the environment to favor the spread of these changes). As people are able to grasp images faster and faster, for example, video and movie cuts become faster still, pushing the boundaries of perception yet further and making the subliminal threshold a moving target. So with information-intensive environments: as young people move further into hyper attention, they modify their environments so that they become yet more information intensive, for example by listening to music while surfing the web and simultaneously writing an essay, a phenomenon that the *Generation M* report found to be increasingly common.

Moreover, it is not only young people who are affected but almost everyone who lives in information-intensive environments. Drawing from the work of neurologist Marc Jeannerod (2002) among others, Catherine Malabou (2008:5) usefully categorizes neural plasticity through three levels: developmental (here synaptogenesis is an important mechanism); synaptic modulation, in which the efficiency of synaptic connections strengthen for those neuronal groups frequently used, while it declines for neuronal groups rarely used; and reparative (as in the case of accident victims or patients with brain lesions). The reparative cases, illuminating for what they reveal about the brain's ability to repurpose and rebuild damaged neuronal networks, are by their nature limited to a relatively few specialized cases. More relevant for our purposes here are the developmental and modulation categories. While the former applies broadly to embryos, infants, and children, the latter shows that neural plasticity remains a vibrant resource for adults as well.

Malabou (2005) notes what has struck many critics (for example, A. Liu [2004] and Galloway and Thacker [2007]), that contemporary models of neuronal functioning, which emphasize networks of neurons rather than a single neuron and that see plasticity as an important, lifelong attribute of brain function and morphology, bear an uncanny resemblance to contemporary global capitalism, which also works through networks and requires continuous rearranging and repurposing of objects and people. (We may note that current models recast in another mode a similar correspondence between neuronal models and techno-economic organization in the late nineteenth century, when nerves and telegraphy were conceptualized in terms of each other, a phenomenon discussed in the next chapter). Malabou's urgent question is this: "*What should we do so that consciousness of the brain does not purely and simply coincide with the spirit of capitalism?*" (2008:12; emphasis in original). Her strategy is to distinguish sharply between flexibility and plasticity; whereas flexibility is all about passive accommodation to the New World Order, plasticity has the potential for resistance and reconfiguration. Becoming conscious of the brain, for her, means becoming conscious of this potential for resistance. But there is a problem with her urgent rhetoric: "becoming conscious" seems to imply that solutions can be found through such high-cognitive functions as decoding written language (e.g., reading her book), but as we have seen, unconscious and nonconscious levels of awareness are also affected (arguably even more than consciousness) by the accelerating pace and "flexibility" demands of global capitalism. How can they be mobilized for resistance? For this, Malabou has no solution, other than to presume that *conscious* (read conceptual) awareness that the brain is plastic will be enough to do the job.

Another possibility, implicit in the concept of technogenesis, is to use digital media to intervene in the cycles of continuous reciprocal causality so that one is not simply passively responding to the pressures of accelerating information flow but using for different ends the very technologies applying this pressure. Malabou argues that a crucial (conceptual) point of intervention should be the gap between neuronal processing and mental awareness (i.e., consciousness), but her argument does not specify how to make this gap work for the goals she purposes. From a technogenetic perspective, the holistic nature of human response to the environment, including conscious, unconscious, and nonconscious awareness, suggests the possibility of fabricating devices that can use unconscious and nonconscious perceptions in ways that make their awareness available to consciousness. An example of such a device is the sociometer developed by Alex Pentland (2008), a professor at MIT, in collaboration with his research group. Capable of sensing subtle movements as well as other unconscious and nonconscious signals, the sociometer creates, in effect, a feedback loop between consciousness and other levels of neuronal responses. Pentland identifies four criteria that can be measured by the sociometer: influence; mimicry, activity, and consistency (2008:4). Because it is difficult or impossible for the conscious mind to monitor these responses, Pentland calls them "honest signals," highly resistant to being faked. His data demonstrate that these signals, detectable by the sociometer, are more reliable in judging outcomes and effects than other (merely) conscious modes of analysis. A similar research program has been initiated at Duke University by my colleague Mark B. N. Hansen, where he and a group of graduate students are developing the somameter, a device that, like the sociometer, can detect unconscious and nonconscious signals such as galvanic skin response and feed that information back to conscious awareness.

Other research programs aim to use feedback between the various levels of conscious, unconscious, and nonconscious responses by creating digital devices that retrain neuronal pathways damaged by accident or illness so that the patient can regain functions crucial to everyday life. Another colleague at Duke University, Miguel Nicolelis, a professor of neuroscience, has developed brain-machine interfaces (which he calls BMIs) that train rats to sense the world magnetically and that teach monkeys to control robotic arms at locations remote from them (Nicolelis 2011). He envisions a future in which paralyzed patients can have their mobility restored by wearing exoskeletons they control by thinking. Still other research programs have used neural plasticity to reorganize brain function so that blind patients can "see" ("Pictured" 2009). These devices work by converting signals from a camera into electrical impulses, which the patient detects through an actuator, about the size of a lollipop, placed on the tongue. In time, the visual cortex is retrained to interpret the electrical impulses as visual images. Other examples include cochlear implants for deaf people; the implants electronically stimulate the inner ear, which the brain can be retrained to interpret as coherent sounds. The practical goals achieved by these research programs vividly demonstrate that plasticity provides not only the grounds for a philosophical call for action but a potent resource for constructive interventions through human-digital media hybridity.

Complex Temporalities in Living and Technical Beings

Now let us circle back and connect these ideas with the earlier discussion of technicity and technical beings. We have seen that in the view of Simondon and others, technical objects are always on the move, temporarily reaching metastability as provisional solutions to conflicting forces. Such solutions are subject to having technical elements detach themselves from a given technical ensemble and becoming reabsorbed in a new technical ensemble, performing a similar or different function there. As Simondon laconically observed, humans cannot mutate in this way. Organs (which he suggested are analogous to technical elements) cannot migrate out of our bodies and become absorbed in new technical ensembles[1] (with the rare exception of organ transplants, a technology that was merely a dream when he wrote his study). Rather than having technical elements migrate, humans mutate epigenetically through changes in the environment, which cause still further epigenetic changes in human biology, especially neural changes in the brain, the central nervous system, and the peripheral nervous system.

Weaving together the strands of the argument so far, I propose that attention is an essential component of technical change (although undertheorized in Simondon's account), for it creates from a background of technical ensembles some aspect of their physical characteristics upon which to focus, thus bringing into existence a new materiality that then becomes the context for technological innovation. Attention is not, however, removed or apart from the technological changes it brings about. Rather, it is engaged in a feedback loop with the technological environment within which it operates through unconscious and nonconscious processes that affect not only the background from which attention selects but also the mechanisms of selection themselves. Thus technical beings and living beings are involved in continuous reciprocal causation in which both groups change together in coordinated and indeed synergistic ways.

We have seen that technical beings embody complex temporalities, and I now want to relate these to the complex temporalities embodied in living beings, focusing on the interfaces between humans and networked and programmable machines. Within a computer, the processor clock acts as a drumbeat that measures out the time for the processes within the machine; this is the speed of the CPU measured in hertz (now megahertz). Nevertheless, because of the hierarchies of code, many information-intensive applications run much slower, taking longer to load, compile, and store. There are thus within the computer multiple temporalities operating at many different time scales. Moreover, as is the case with technical objects generally, computer code carries the past with it in the form of low-level routines that continue to be carried over from old applications to new updates, without anyone ever going back and readjusting them (this was, of course, the cause of the feared Y2K crisis). At the same time, code is also written with a view to changes likely to happen in the next cycle of technological innovation, as a hedge against premature obsolescence, just as new code is written with a view toward making it backward-compatible. In this sense too, the computer instantiates multiple, interacting, and complex temporalities, from microsecond processes up to perceptible delays.

Humans too embody multiple temporalities. The time is takes for a neuron to fire is about 0.3 to 0.5 milliseconds. The time it takes for a sensation to register in the brain ranges from a low estimate of 80 milliseconds (Pockett 2002) to 500 milliseconds (Libet et al. 1979), or 150 to 1,000 times slower than neural firing. The time it takes the brain to grasp and understand a high-level cognitive facility like recognizing a word is 200–250 milliseconds (Larson 2004), or about six times slower than the lower estimate for registering a sensation. Understanding a narrative can of course take anywhere from several minutes to several hours. These events can be seen as a linear sequence—firing, sensation, recognizing, understanding—but they are frequently not that tidy, often happening simultaneously and at different rates concurrently. Relative to the faster processes, consciousness is always belated, reaching insights that emerge through enfolded time scales and at diverse locations within the brain. For Daniel Dennett (1991), this means the consistent and unitary fabric of thought that we think we experience is a confabulation, a smoothing over of the different temporalities that neural processes embody. Hence his "multiple drafts" model, in which different processes result in different conclusions at different times.

Now let us suppose that the complex temporalities inherent in human cognitive processing are put into continuous reciprocal causation with machines that also embody complex temporalities. What is the result? Just as some neural processes happen much faster than perception and much, much faster than conscious experience, so processes within the machine happen on time scales much faster even than neural processes and far beyond the threshold of human perception. Other processes built on top of these, however, may not be so fast, just as processes that build on the firing of neural synapses within the human body may be much slower than the firings themselves. The point at which computer processes become perceptible is certainly not a single value; subliminal perception and adaptive unconsciousness play roles in our interactions with the computer, along with conscious experience. What we know is that our experiences with the diverse temporalities of the computer are pushing us toward faster response times and, as a side effect, increased impatience with longer wait times, during which we are increasingly likely to switch to other computer processes such as surfing, checking e-mail, playing a game, etc. To a greater or lesser extent, we are all moving toward the hyper attention end of the spectrum, some faster than others.

Going along with the feedback loops between the individual user and networked and programmable machines are cycles of technical innovation. The demand for increased information-intensive environments (along with other market forces such as competition between different providers) is driving technological innovations faster and faster, which can be understood in Simondon's terms as creating a background of unrealized potential solutions (because of a low degree of technicity, that is, unresolved conflicts between different technical forces and requirements). Beta versions are now often final versions. Rather than debugging programs completely, providers rush them to market and rely on patches and later versions to fix problems. Similarly, the detailed documentation meticulously provided for programs in the 1970s and 1980s is a thing of the past; present users rely on help lines and lists of "known problems" and fixes. The unresolved background created by these practices may be seen as the technical equivalent to hyper attention, which is both produced by and helps to produce the cycles of technical innovation that result in faster and faster changes, all moving in the direction of increasing the information density of the environment.

This, then, is the context within which Steve Tomasula's electronic multimodal novel *TOC* was created. *TOC* is, in a term I have used elsewhere, a technotext (Hayles 2002): it embodies in its material instantiation the complex temporalities that also constitute the major themes of its narrations. In this sense it is analogous to the digital feedback devices discussed earlier, for it evokes a range of conscious, unconscious, and nonconscious responses that manifest through affect as well as attention, unconscious awareness as well as conscious analysis of themes and motifs. In Malabou's terms, it (along with many literary texts) can be seen as a device for evoking responses at all levels of engagement and, through reflexive feedback loops that connect affect to conceptualization, bring them into metacognitive awareness, which (we recall) Bargh suggested is "the workplace where one can assemble and combines the various components of complex percpetual-motor skills" (2005:53). Heterogeneous in form, bearing the marks of ruptures created when some collaborators left the scene and others arrived, *TOC* can be understood as a metonym for the gap between the neuronal protoself and the narrated self. It explores the relation between human bodies and the creation and development of networked and programmable machines, with both living and technical beings instantiating and embodying complex temporalities that refuse to be smoothly integrated into a rational and unitary scheme of a clock ticking. It thus simultaneously testifies to and resists the "spirit of capitalism" in the era of globalization that Catherine Malabou urges is the crucial problem of our time—and our temporalities.

Modeling *TOC*

As a multimodal electronic novel, *TOC* offers a variety of interfaces, each of which has its own mode of pacing. The video segments, with voiceover and animated graphics, do not permit interaction by the user (other than to close them) and so proceed at a preset pace. The textual fragments, such as those that appear in the bell jar graphic, are controlled by the user scrolling down and so have a variable, user-controlled pace. Access to the bell jar fragments is through a player-piano interface, which combines user-controlled and preset paces: the scrolling proceeds at a constant rate, but the user chooses when to center the crosshair over a "hole" and click, activating the link that brings into view bell jar text fragments, brief videos, and cosmic datelines.

Through its interfaces, *TOC* offers a variety of temporal regimes, a spectrum of possibilities enacted in different ways in its content.

An early section, a thirty-three-minute video centering on a *Vogue* model, encapsulates competing and cooperating temporalities within a narrative frame. The narration opens and closes with the same sentence, creating a circular structure for the linear story within: "Upon a time, a distance that marked the reader's comfortable distance from it, a calamity befell the good people of X." The calamity is nothing more (or less) than them sharing the same present for a moment before falling into different temporal modes. Their fate is personified in the unnamed model, ripped from her daily routine when her husband engages "in a revelry that ended in horrible accident." A car crash lands him into immediate surgery, which ends with him in a coma, kept alive by a respirator breathing air into his body and a pump circulating his blood, thus transforming him into an "organic machine." Instead of hectic photo shoots and an adrenaline-driven lifestyle, the model now spends uncounted hours at his bedside, a mode of life that makes her realize the arbitrariness of time, which "existed only in its versions." She now sees the divisions into hours, minutes, and seconds as an artificial grid imposed on a holistic reality, which the graphics visually liken to the spatial grid Renaissance artists used to create perspectival paintings.

In this image (fig. 4.1), the basic homology is between clock time and spatial perspective, a spatialization of temporality that converts it into something that can be measured and quantified into identical, divisible and reproducible units. The screen shot illustrates the video's design aesthetic. The painter's model, a synecdoche for the protagonist, is subjected to a spatial regime analogous to the temporal regime the *Vogue* model followed in her hectic days. The large concentric rings on the left side are reminiscent of a clock mechanism, visually reinforcing the narrative's juxtaposition of the 360 seconds in an hour and the 360 degrees in a circle, another spatial-temporal conjunction. These circular forms, repeated by smaller forms on the right, are cut by the strong horizontal line of the painter and his model, along with the text underneath that repeats a phrase from the voiceover, "Tomorrow was another day," a banal remark that the model takes to represent the spatialization of time in general. The circular repetitions combined with a horizontal through line visually reproduce the linguistic form of the narration, which proceeds as a linear sequence encapsulated within the opening and closing repetitions, which also appear in the narrative in various syntactic permutations.

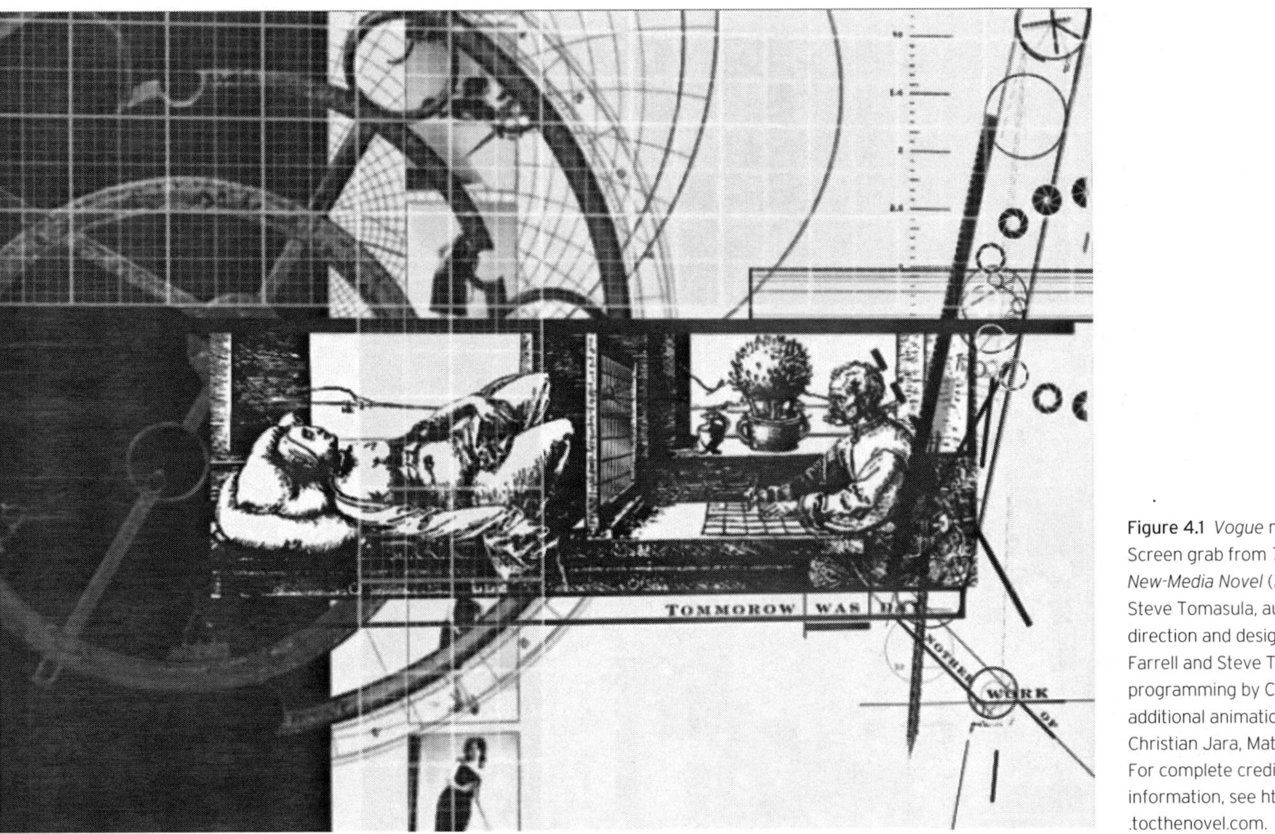

Figure 4.1 *Vogue* model video. Screen grab from *TOC: A New-Media Novel* (2009) by Steve Tomasula, author, with direction and design by Stephen Farrell and Steve Tomasula; programming by Christian Jara; additional animation and art by Christian Jara, Matt Lavoy, et al. For complete credits and more information, see http://www.tocthenovel.com.

When the doctors inform the model that "she could decide whether or not to shut her husband [i.e., his life support] off," she begins contemplating a "defining moment, fixed in consciousness" that will give her life a signifying trajectory and a final meaning. She (and the animated graphics) visualize her looming decision first as a photograph, in which she sees her husband's face "like a frozen moment," unchanging in its enormity akin to a shot of the Grand Canyon, and then as two films in different temporal modes: "his coiled tight on the shelf, while hers was still running through a projector." Her decision is complicated when the narration, in a clinamen that veers into a major swerve, announces, "The pregnancy wasn't wanted."

We had earlier learned that, in despair over her husband's unchanging condition, she had begun an affair with a "man of scientific bent." Sitting by her husband's bedside, she wonders what will happen if she cannot work and pay "her husband's cosmic bills" when the pregnancy changes her shape and ends her modeling career, already jeopardized by her age compared to younger and younger models. As she considers aborting, the graphics and narration construct another homology: pull the plug on her husband and/or cut the cord on the "growth" within. While her husband could go on indefinitely, a condition that wrenched her from measured time, the fetus grows according to a biological time line with predictable phases and a clear endpoint. The comparison causes the metaphors to shift: now she sees her predicament not so much as the arrival of a "defining moment" as a "series of other moments" that "suggested narrative." Unlike a single moment, narrative (especially in her case) will be defined by the ending, with preceding events taking shape in relation to the climax: "The end would bestow on the whole duration and meaning."

The full scope of her predicament becomes apparent when the narration, seemingly offhandedly, reveals that her lover is her twin brother. In a strange leap of logic, the model has decided that "by sleeping with someone who nominally had the same chromosomes," she really is only masturbating and not committing adultery. Admitting that she is not on an Olympian height but merely "mortal," a phrase that in context carries the connotation of desire, she wonders, if time exists only in its versions, "why should she let time come between her brother and her mortality." Her pregnancy is, of course, a material refutation of her fantasy. Now, faced with the dual decisions of pulling the plug and/or cutting the cord, she sees time as an inexorable progression that goes in only one direction. This inexorability is contested in the narrative by the perverse wrenching of logical syllogisms from the order they should ideally follow, as if in reaction against temporal inevitability in any of its guises. The distortion appears repeatedly, as if the model is determined not to agree that A should be followed by B. For example, when she asks her lover/brother why time goes in only one direction, he says (tautologically) that "disorder increases in time because we measure time in the direction that disorder increases." In response, she thinks "that makes it possible for a sister to have intercourse with her brother," a non sequitur that subjects the logic of the tautology to an unpredictable and illogical swerve.

As much as the model tries to pry time from sequentiality, however, it keeps inexorably returning. To illustrate his point about the second law of thermodynamics giving time its arrow, the lover/brother reinstalls temporal ordering by pointing out that an egg, being broken, never spontaneously goes back together. While the graphics show an egg shattering—an image recalling the fertilization of the egg that has resulted in the "growth" within—he elaborates by saying it is a matter of probability: while there is only one arrangement in which the egg is whole, there are infinite numbers of arrangements in which it shatters into fragments. Moreover, he points out that temporality in a scientific sense manifests itself in three complementary versions: the thermodynamic version, which gives time its arrow; the psychological version, which reveals itself in our ability to remember the past but not the future; and the cosmological version, in which the universe is expanding, creating the temporal regime in which humans exist.

Faced with this triple whammy, the model stops trying to resist the ongoingness of time and returns to her earlier idea of time as the sequentiality of a story, insisting that "a person had to have the whole of narration" to understand its meaning. With the ending, she thinks, the story's significance will become clear, although she recognizes that this hope involves "a sticky trick that depended on a shift of a tense," that is, from present, when everything is murky, to past, when the story ends and its shape is fully apparent. The graphics at this point change from the iconic and symbolic images that defined the earlier aesthetic to the indexical correlation of pages falling down, as if a book were being riffled through from beginning to end. However, the images gainsay the model's hope for a final definitive shape, for as the pages fall, we see that they are ripped from heterogeneous contexts: manuscript pages, typed sheets, a printed bibliography with lines crossed out in pen, an index, a child's note that begins "Dear Mom," pages of stenographic shapes corresponding to typed phrases. As if in response to the images, the model feels that "a chasm of incompletion opened beneath her." She seems to realize that she may never achieve "a sense of the whole that was much easier to fake in art than in life."

With that, the narrative refuses to "fake" it, opting to leave the model on the night before she must make a decision one way or another, closing its circle but leaving open all the important questions as it repeats the opening sentence: "Upon a time, a distance that marked the reader's comfortable distance from it, a calamity befell the good people of X." We hear the same sentence, but in another sense we hear an entirely different one, for the temporal unfolding of the narrative has taught us to parse it in a different way. "Upon a time" differs significantly from the traditional phrase "Once upon a time," gesturing toward the repetitions of the sentence that help to structure the narrative. The juxtaposition of time and distance ("upon a time, a distance") recalls the spatialization of time that the model understood as the necessary prerequisite to the ordering of time as a predictable sequence. The "comfortable distance" that the fairy-tale formulation supposedly creates between reader and narrative is no longer so comfortable, for the chasm that opened beneath the model yawns beneath us as we begin to suspect that the work's "final shape" is a tease, a gesture made repeatedly in *TOC*'s different sections but postponed indefinitely, leaving us caught between different temporal regimes.

The "calamity [that] befell the good people of X," we realize retrospectively after we have read further, spreads across the entire work, tying together the construction of time with the invention of the "Difference Engine," also significantly called the "Influencing Machine." The themes introduced in the video sequence continue to reverberate through subsequent sections: the lone visionary who struggles to capture the protean shapes of time in narrative; the contrast between time as clock sequence and time as a tsunami that crashes through the boundaries of ordered sequence; the transformations time undergoes when bodies become "organic machines"; the mechanization of time, a technological development that conflates the rhythms of living bodies with technical beings; the deep interrelation between a culture's construction of time and its socius; and the epistemic breaks that fracture societies when they are ripped from one temporal regime and plunged into another.

Capturing Time

Central to attempts to capture time is spatialization. When time is measured, it is frequently through spatialized practices that allow temporal progress to be visualized—hands moving around an analogue clock face, sand running through an hourglass, a sundial shadow that moves as the sun crosses the sky. In the "Logos" section of *TOC*, the spatialization of time exists in whimsical as well as mythic narratives. The user enters this section by using the cursor to "move" a pebble into a box marked "Logos" (the other choice is "Chronos," which leads to the video discussed above), whereupon the image of a player-piano roll begins moving, accompanied by player-piano-type music (fig. 4.2). Using a crosshair, the user can click on one of the slots: blue for narrative, red for short videos, green for distances to the suns of other planetary systems. The order in which a user clicks on the blue slots determines which narrative fragments can be opened; different sequences lead to different chunks of text being made available.

The narratives are imaged as text scrolls on old paper encapsulated within a bell jar frame, as if to emphasize their existence as archival remnants. The scrolls pick up on themes already introduced in the "Chronos" section, particularly conflicting views of time and their relation to human-machine hybrids. The difficulty of capturing time as a thing in itself is a pervasive theme, as is the spatialization of time. For example, capturing the past is metaphorically rendered as a man digging a hole so deep that his responses to questions from the surface come slower and slower, thus opening a gap between the present of his interlocutors and his answers, as if he were fading into the past. Finally he digs so deep that no one can hear him at all, an analogy to a past that slips into oblivion (fig. 4.3).

The metaphor for the future is a woman who climbs a ladder so tall that the details of the surface appear increasingly small, as if seen from a future perspective where only large-scale trends can be projected. As she climbs higher, family members slip into invisibility, then her village, then the entire area. In both cases, time is rendered as spatial movement up or down. The present, it seems, exists only at ground level.

The distinction between measured time and time as temporal process can be envisioned as the difference between exterior spatialization and interior experience: hands move on a clock, but (as Bergson noted) heartbeat, respiration, and digestion are subjective experiences that render time as duration and, through cognitive processes, as memory. The movement from measured time to processual temporality is explored through a series of fragments linked to the Difference Engine. The allusion, of course, is to Charles Babbage's device, often seen as the precursor to the modern computer (more properly, the direct precursor is his Analytical Engine). In "Origins of the Difference Engine, 1," for example, the narrative tells of "a woman who claimed she had invented a device that could store time." Her device is nothing other than a bowl of water. When she attempts to demonstrate it

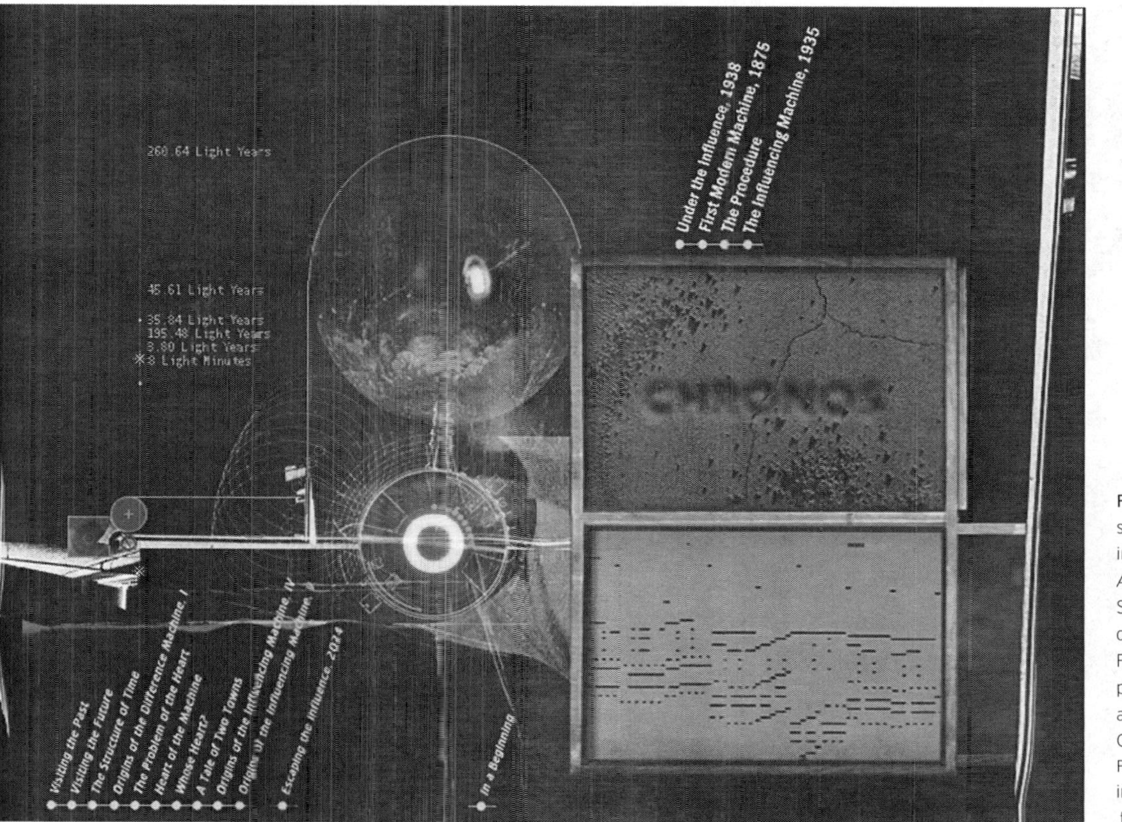

Figure 4.2 "Chronos/Logos" section showing player-piano interface. Screen grab from *TOC: A New-Media Novel* (2009) by Steve Tomasula, author, with direction and design by Stephen Farrell and Steve Tomasula; programming by Christian Jara; additional animation and art by Christian Jara, Matt Lavoy, et al. For complete credits and more information, see http://www.tocthenovel.com.

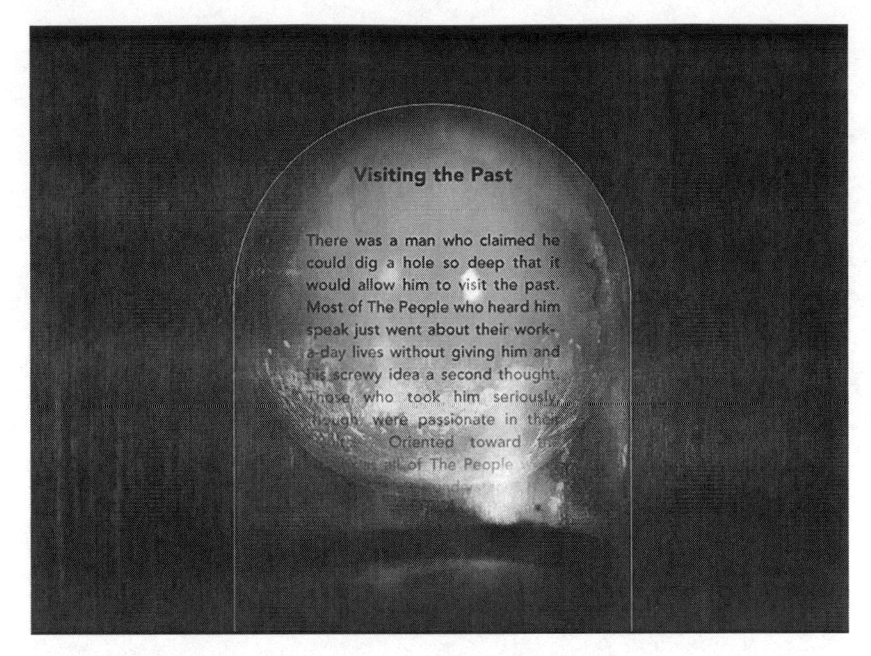

Figure 4.3 Bell jar graphic with "Visiting the Past" narrative. Screen grab from *TOC: A New-Media Novel* (2009) by Steve Tomasula, author, with direction and design by Stephen Farrell and Steve Tomasula; programming by Christian Jara; additional animation and art by Christian Jara, Matt Lavoy, et al. For complete credits and more information, see http://www.tocthenovel.com.

to the townspeople, who have eagerly gathered because they can all think of compelling reasons to store time, they laugh at her simplicity. To prove her claim, she punches a hole in the bowl, whereupon the water runs out. The running water is a temporal process, but the bowl can be seen as storing this process only so long as the process itself does not take place. The paradox points to a deeper realization: time as an abstraction can be manipulated, measured, and quantified, as in a mathematical equation; time as a process is more protean, associated with Bergson's duration rather than time in its spatialized aspects.

In a series of narrative fragments the bowl of water continues to appear, associated with a linage of visionaries who struggle to connect time as measurement with time as process. In "Heart of the Machine," a woman who first tried to stop time as a way to preserve her youth imperceptibly becomes an old hermit in the process. As she ages, she continues to grapple with "the paradox of only being able to save time when one was out of time." She glances at the mattress and springs covering her window—an early attempt to insulate the bowl of water from outside influences—and then thinks of the water running as itself a spring. Thereupon "she realized that the best humans would ever be able to do, that the most influential machine they could ever make would actually be a difference engine—a machine, or gadget, or engine of some sort that could display these differences and yet on its face make the vast differences between spring and springs—and indeed all puns—as seeming real as any good illusion." The Difference Engine, that is, is a "gadget" that can operate like the Freudian unconscious, combining disparate ideas through puns and metonymic juxtapositions to create the illusion of a consistent, unitary reality. The computer as a logic machine has here been interpreted with poetic license as a machine that can somehow combine differences—preeminently the difference between measured time and processual temporality—into a logos capable of influencing entire human societies.

What could possibly power such an exteriorized dream machine? The woman has a final intuition when "at that very instant, she felt a lurch in the lub-dub of her heart, confirming for her the one course that could power such a device, that could make time's measurements, its waste and thrift not only possible but seemingly natural." That power source is, of course, the human heart. Subsequent narratives reveal that her heart has indeed been encapsulated within the Influencing Machine, the timekeeping (better, the time-constructing) device that subsequently would rule society. The dream logic instantiated here conflates the pervasiveness and ubiquity of contemporary networked and programmable machines that actually do influence societies around the globe, with a steampunk machine that has as its center a human heart, an image that combines the spatialization of time with the duration Bergson posed as its inexplicable contrary. Such conflation defies ordinary logic in much the same way as the *Vogue* model's non sequiturs defy the constructions of time in scientific contexts. Concealed within its illogic, however, is a powerful insight: humans construct time through measuring devices, but these measuring devices also construct humans through their regulation of temporal processes. The resulting human-technical hybridization in effect conflates spatialized time with temporal duration (fig. 4.4).

The feedback loop reaches its apotheosis in a video segment that begins, "When it came, the end was catastrophic," accessed by clicking on a cloudy section in the piano roll. The video relates an era when the triumph of the Influencing Machine seems absolute as it produces measured and spatialized time; "the people had become so dependent on time that when they aged, it permeated their cells." The animated graphics extend the theme,

Figure 4.4 The Influencing Machine just before it breaks (at 12:00:00). Screen grab from *TOC: A New-Media Novel* (2009) by Steve Tomasula, author, with direction and design by Stephen Farrell and Steve Tomasula; programming by Christian Jara; additional animation and art by Christian Jara, Matt Lavoy, et al. For complete credits and more information, see http://www.tocthenovel.com.

showing biological parts—bones, egg-like shapes, mobile microscopic entities that seem like sperm—moving within the narrative frame of time measured. Spatialized time also permeates all the machines, so that everything becomes a function of the "prime Difference Engine," which "though it regulated all reality could not regulate itself." Like the historical Difference Engine Babbage created, the Influencing Machine has no way to self-correct, so that "minute errors accumulated" until one day, "like a house of cards, time collapsed." Like the bowl of water associated with the female mystic's humiliation but now magnified to a vast scale, the transformation from time measured to processual time has catastrophic effects: "Archeologists later estimated that entire villages were swept away, because they had been established at the base of dikes whose builders had never once considered that time could break."

As the city burns and Ephema, queen of the city, escapes with her followers, her "womb-shaped tears" as she weeps for her destroyed city "swelled with pregnancy," presumably the twins Chronos and Logos, who an earlier video explains were born in different time zones, rendering ambiguous who was born first and has the right of primogeniture. The father, the people speculate, must be none other than the Difference Engine itself, an inheritance hinting that the struggle between measured time and processual time has not ended. Indeed, when the queen establishes a new society, her hourglass fingernails spread throughout the populace, so that soon all the people have hourglass fingernails. During the day, "the sand within each fingernail ran toward their fingertips." To balance this movement of time, the people must sleep with their arms in slings so that the sand runs back toward their palms. Nevertheless, "it was only a matter of time until a person was caught with too much sand in the wrong places," whereupon he dies. This conflation of time as experience and time as measured, in another version of human-technical hybridity, is given the imprimatur of reality, for such a death is recorded as resulting from "Natural Causes," a conclusion reinforced by a graphic of the phrase appearing in copperplate handwriting. The episode recalls Bruno Latour's comment in *Pandora's Hope: Essays on the Reality of Science Studies* (1999) that in the contemporary period, "time enmeshes, at an ever greater level of intimacy and on an ever greater scale, humans and nonhumans with each other. . . . The confusion of humans and nonhumans is not only our past *but our future as well*" (200; emphasis in original).

In *TOC*, time is always on the move, so that no conclusion or temporal regime can ever be definitive. In a text scroll entitled "The Structure of Time," the narrative locates itself at "the point in history when total victory seemed to have been won by the Influencing Machine and its devotees, when all men, women, and thought itself seemed to wear uniforms." An unnamed female protagonist (perhaps the mystic whose heart powers the Influencing Machine) envisions the Influencing Machine as "an immense tower of cards," a phrase that echoes the collapse of time "like a house of cards." The Influencing Machine is usually interpreted as a vertical structure in which each layer builds upon and extends the layer below in an archaeology of progress—sundials giving way to mechanical clocks above them, digital clocks higher still, atomic clocks, then molecules "arranged like microscopic engines" (nanotechnology?), and finally, above everything, "a single atom" moving far too fast for human perception. The protagonist, however, in a moment of visionary insight inverts the hierarchy, so that each layer becomes "subservient to the layers below, and could only remain in place by the grace of inertia." Now it is not the increasing acceleration and sophistication that counts but the stability of the foundation. All the clocks rest on the earth, and below that lies "the cosmos of space itself," "the one layer that was at rest, the one layer against which all other change was measured and which was, therefore, beyond the influence of any machine no matter what kind of heart it possessed, that is, the stratum that made all time possible—the cold, infinite vastness of space itself." Echoing the cosmological regime of time that the *Vogue* model's brother says makes the other temporal regimes possible, the mystic's vision reinstates as an open possibility a radically different temporal regime. In effect, what the mystic has done is invert the pressures of information-intensive environments so that now it is not their velocity that counts but the cosmological stability on which they ultimately depend. Temporal complexity works in *TOC* like a Möbius strip, with one temporal regime transforming into another just as the outer limit of the boundary being traced seems to reach its apotheosis.

This dynamic implies that the Influencing Machine will also reinstate itself just when the temporality of the cosmos seems to have triumphed. So the video beginning with the phrase "When it came, the end was catastrophic" concludes with the emergence of The Island, which I like to think of as The Island of Time.

Time and the Socius

Divided into three social (and temporal) zones, The Island explores the relation between temporal regimes and the socius (fig. 4.5). The left hand (or western) side believes in an ideology of the past, privileging practices and rituals that regard the present and future either as nonexistent or unimportant. The middle segment does the same with the present, and the right or easternmost part of The Island worships the future, with its practices and rituals being mirror inversions of the westernmost portion. Each society regards itself as the true humans—"the Toc, a name that meant The People"—while the others are relegated to the Tic, nonpeople excluded from the realm of the human. Echoing in a satiric vein the implications of the other sections, The Island illuminates the work's title, *TOC*. "TOC" is a provocation, situated after a silent "Tic," like a present that succeeds a past; it may also be seen as articulated before the subsequent "Tic," like a present about to be superseded by the future. In either case, "TOC" is not a static entity but a provisional stability, a momentary pause before (or after) the other shoe falls. Whereas Simondon's theory of technics has much to say about nonhuman technical beings, *TOC* has more to reveal about human desires, metaphors, and socius than it does about technical objects. Often

Figure 4.5 "The Island." Screen grab from *TOC: A New-Media Novel* (2009) by Steve Tomasula, author, with direction and design by Stephen Farrell and Steve Tomasula; programming by Christian Jara; additional animation and art by Christian Jara, Matt Lavoy, et al. For complete credits and more information, see http://www.tocthenovel.com.

treating technical objects as metaphors, *TOC* needs a theory of technics as a necessary supplement, even as it also instantiates insights that contribute to our understanding of the coconstitution of living and technical beings.

In conclusion, I turn briefly to consider *TOC*'s own existence as a technical object. An aspect of *TOC* likely to strike most users immediately is its heterogeneity, by which I mean not only the multiple modalities of its performances but also its diverse aesthetics. The credits, which roll after a user has completed reading the segments of The Island, reveal the many collaborators that Steve Tomasula had in this project. Lacking the budget that even a modest film would have and requiring expertise across a wide range of technical skills, *TOC* is something of a patchwork, with different collaborators influencing the work in different directions at different times. Its aesthetic heterogeneity points to the difficulties writers face when they attempt to move from a print medium, in which they operate either as sole author or with one or two close collaborators, to multimodal projects involving expertise in programming, music and sound, animation, graphic design, visual images, and verbal narration, to name some of the skills required. Lacking big-time financing, literary authors are forced to stretch the project across years of effort—as in the case of *TOC*'s composition—relying on the good offices of collaborators who will do what they can when they can.

TOC gestures toward a new regime, then, in which artistic creation is happening under very different conditions than for print literature. It would be possible to approach the work as if it were an inferior piece of print literature, marred by its lack of unity and inconsistent artistic design. I prefer to situate it in a context that Simondon associates with virtuality—a work that is more abstract than concretized and that consequently possesses a large reservoir of possibilities that may be actualized as the technological milieu progresses and as technical objects proliferate, exploring ways to achieve greater integration. Or perhaps not. The accelerating pace of technological change may indicate that traditional criteria of literary excellence are very much tied to the print medium as a mature technology that produces objects with a large degree of concretization. In newer technical milieu, changing so fast that a generation may consist of only two or three years, the provisional metastability of technical individuals may become even less stable, so that it is more accurate to speak of continuous transformation than metastability at all.

TOC anticipates this by revealing the inherent instability of temporal regimes, the cataclysmic breaks that occur when societies are wrenched from one regime and plunged into another, and the inability of the narratives to create any temporal regime that will be both durable and all-encompassing. In this, it may be viewed as commenting on its own conditions of possibility. The circularity of a work that both produces and is produced by its aesthetic possibilities may be a better criterion to judge its achievement than the relatively settled conventions of print. Like the model of cognitive-technical transformation proposed herein, *TOC* is inhabited by complex temporalities that refuse to resolve into clear-cut sequences or provide definitive boundaries between living and technical beings.

5

Technogenesis in Action

Telegraph Code Books and the Place of the Human

As we have seen, contemporary technogenesis implies continuous reciprocal causality between human bodies and technics. Frequently these cycles have the effect of reengineering environments so as to favor further changes. Flowing out into the wider society, they catalyze profound cultural, economic, and political transformations. Chapters 2 and 3 show how the technogenetic spiral changes brain morphology and function (in the case of reading on the web) and disciplinary assumptions and practices (in the case of the Digital Humanities). Notwithstanding that they help to make the case for transformative change, the range of examples in these chapters is limited to specific activities and sites. A case study that brings in more of the connections between epigenetic changes in human biology, technological innovations, cultural imaginaries, and linguistic, social, and economic changes would be useful.

My candidate is the first globally pervasive binary signaling system, the telegraph. Telegraphy is useful in part because it is a relatively simple technology compared to the digital computer, yet in many ways it anticipated the far-reaching changes brought about by the Internet (Standage 2007). Moreover, the telegraph's transformative effects on nineteenth-century practices, rhetoric, and thought are well documented (Carey 1989; Peters 2006; Otis 2001; Menke 2008). Extending these analyses, my focus in this chapter is on the inscription technology that grew parasitically alongside the monopolistic pricing strategies of telegraph companies: telegraph code books.[1] Constructed under the bywords "economy," "secrecy," and "simplicity," telegraph code books matched phrases and words with code letters or numbers.[2] The idea was to use a single code word instead of an entire phrase, thus saving money by serving as an information compression technology. Generally economy won out over secrecy, but in specialized cases, secrecy was also important.

Although telegraph code books are now obsolete, a few hundred (of the hundreds of thousands originally published) have been preserved in special library collections and the code book collecting community. These remnants of a once flourishing industry reveal the subtle ways in which code books affected assumptions and practices during the hundred years they were in use. These effects may be parsed through three sets of dynamic interactions: bodies and information; code and language; and messages and the technological unconscious. The dynamic between bodies and information reveals the tension between visions of frictionless communication and the stubborn materialities of physical entities. With the advent of telegraphy, messages and bodies traveled at unprecedented speeds by wire and rail. This regime of speed, crucial to telegraphy's reconfiguration of cultural, social, and economic environments, led to troubled minglings of bodies and messages, as if messages could become bodies and bodies messages. At the same time, telegraphy was extraordinarily vulnerable to the resistant materialities of physically embodied communication, with constant breakdown of instruments and transmission lines and persistent human error. In this sense telegraphy was prologue to the ideological and material struggle between dematerialized information and resistant bodies characteristic of the globalization era, a tension explored below in the section on information and the cultural imaginaries of bodies.

The interaction between code and language shows a steady movement away from a human-centric view of code toward a machine-centric view, thus anticipating the development of full-fledged machine codes with the digital computer. The progressive steps that constituted this shift reveal the ways in which monopoly capital, technical innovations, social conditions, and cultural imaginaries entwined to bring about significant shifts in conscious and unconscious assumptions about the place of the human. As James Carey (1989) has brilliantly demonstrated, the decoupling of message transmission from physical modes of transport initiated regional, national, and global transformations that catalyzed even more increases in the speed of message transmissions. The line of descent that runs from telegraphy through the Teletype and to the digital computer connects these environmental changes with the epigenetic transformations in human cognitive capacities discussed in chapters 2, 3, and 4.

Like oozing tree sap that entraps insects and hardens into amber, the code books captured and preserved telegraphic communications from the mid-nineteenth to the mid-twentieth century, principally in the United States and Europe. The phrases and words contained in code books represent messages that the compilers thought were likely to be useful for a wide variety of industries and social situations. Virtually every major industry had code books dedicated to its needs, including banks, railroads, stock trading, hospitality, cotton, iron, shipping, rubber, and a host of others, as well as the military. They reveal changing assumptions about everyday life that the telegraph initiated as people began to realize that fast message transmission implied different rhythms and modes of life. Thus the telegraph code books bear witness to shifts in the technological unconscious that telegraphy brought about.

Along with the invention of telegraphic codes comes a paradox that John Guillory has noted: code can be used both to clarify and occlude (2010b). Among the sedimented structures in the technological unconscious is the dream of a universal language. Uniting the world in networks of communication that flashed faster than ever before, telegraphy was particularly suited to the idea that intercultural communication could become almost effortless. In this utopian vision, the effects of continuous reciprocal causality expand to global proportions capable of radically transforming the conditions of human life. That these dreams were never realized seems, in retrospect, inevitable. Their significance for this book lies less in the cultural imaginary they articulate than in their anticipation of the tensions between narrative and database that will be the focus of the subsequent group of chapters.

Bodies and Information

Like Janus, nineteenth-century telegraphy had two faces. On the one hand, writers enthusiastically embraced it as a medium akin to spirituality in its mystery, its "annihilation of time and space,"[3] and its ability to convey thoughts faster than the wind. Edward Bryn (1900) is typical in his claim that "through its agency thought is extended beyond the limitations of time and space, and flashes through air and sea around the world" (15). On the other hand, telegraphy's formative years constitute a saga of resistant materiality, frustratingly inert and maddeningly uncooperative. Cables that repeatedly broke under tension, underground conduits that went bad, poles that rotted, glass insulators that were defective or shot out by vandals, gutta-percha insulation that failed in riverbeds, wires that were either too heavy to hold, or if thinner, were too fragile and had resistances too high for effective transmission—these and a hundred other ills made telegraphy a mode of communication uncommonly bound precisely to the exigencies of time and space (Thompson 1947). Time and space were not, common wisdom to the contrary, annihilated by the telegraph, but they were reconfigured. The reconfiguration had the effect of entangling monopoly capitalism with the new technology so that it was no longer possible for capital to operate without the telegraph or its successors; nor was it possible, after about 1866, to think about the telegraph without thinking about monopoly capital.

In the intervening period, the chaotic jumble of different telegraph companies during 1840–66, each with its own protocols and ambitions that often included conspiracies to thwart business rivals, added to the confusion. Given these conditions, the wonder is that messages got through at all, as Robert L. Thompson's magisterial history *Wiring a Continent: The History of the Telegraph Industry in the United States, 1832–1866* (1947) makes clear. Marshall Lefferts of the New York and New England, or Merchants' Telegraph line, sketched the situation in a stockholders report in 1851:

> A presents himself to our Boston office to send a message . . . to New Orleans. We receive and send it to New York, and there hand it over to one of the Southern lines, paying them at the same time the price of transmission for the whole distance, we simply deducting for our service performed. And so the message is passed on, either to stop on the way, or by good luck to reach its destination. If it does not reach its destination . . . [we tell the sender] we will make inquiries, and if we can learn which line is at fault we will return him his money. We make inquiry, and when I tell you we can get no satisfaction, it is almost the universal answer; for they all insist on having sent the message through. (qtd. in Thompson 1947:251)

Like the various players involved in the massive Gulf Coast oil spill who protested their own innocence and pointed fingers at the others, the competing companies protested they had sent the messages through correctly and someone else must be to blame. Message transmission was thus dependent on multiple functionalities, most of which lay outside the control—or even the knowledge—of the individual consumer. In this sense, telegraphy anticipated the message transmission system of the Internet, in which an individual user may have no knowledge or awareness of the complex pathways that information packets take en route to their destinations.

As we know, the decision to standardize the protocols and not the instruments, companies, or transmission lines has been crucial to the (relatively) seamless transmission of Internet messages. Unlike the Internet, telegraphy depended on a wide variety of transmission protocols associated with the distinctive technologies of different instruments. If telegraphy was like Janus in having two faces, it was like Proteus in assuming dozens of different bodies. Leaving aside optical and semaphoric telegraphs, electric telegraphs alone took an astonishing profusion of forms (Gitelman 1999; Thompson 1947). As a former chemist, I am particularly charmed by an early version that used electrolytic reactions; an electric battery sent signals, which were transmitted at the receiving end to parallel wires immersed in tubes filled with acid solution, which released hydrogen bubbles when a current passed through the wires (Otis 2001; Thompson 1947:4). The Cooke and Wheatstone used swinging magnetic needles placed at the receiving ends of wires through which a current passed. The House printing telegraph (discussed below) used a keyboard reminiscent of a piano, with each key standing for a letter of the alphabet. Even the sturdy and relatively simple Morse machine underwent considerable modification before assuming its final form.

After about 1850, the Morse instrument began to establish a steady superiority over its rivals, largely because it was reliable in a wide variety of conditions, including faulty lines and high resistances. Morse code was also standardized, with defined time intervals distinguishing between the different symbols and spaces. Since space could not be conveyed directly in telegraphy (a time-based medium), the spatial separation between letters and words in printed texts were translated into temporal equivalents. One dit interval indicated the space between each dit and dah, three dit intervals indicated a spacing between letters, and seven dit intervals indicated spacing between words. This temporal regime required exact timing on the operator's part. Indeed, the main development from the 1840s to the 1860s was not technological but physiological: operators discovered they could translate the signals directly, without the necessity of copying down dits and dahs and having a copyist translate the code into letters.

Although reluctant to abandon the printed record provided by the tape, telegraph officials were persuaded by the savings in time and money to adopt the new system. Shortly thereafter the "sounder" was invented, facilitating sound reception by amplifying the signals into short (dit) and long (dah) bursts of sound. Tal. B. Shaffner, writing in 1859, remarked upon the change: "Some years ago, as president of a telegraph line, I adopted a rule forbidding the receiving of messages by sound. Since then the rule has been reversed, and the operator is required to receive by sound or he cannot get employment in a first class station" (qtd. in Thompson 1947:250). The learning curve for receiving by sound is at least as steep as learning to touch-type. Whereas most typing manuals indicate that touch-typing can be learned in a month and that speeds in excess of 90 words per minute are not unusual, sound receiving had a typical proficiency range of about 40 code groups (i.e., words) per minute. The skill was quantified by the US Army in 1945; after forty hours practice in sound receiving, 99 percent of students could transcribe at 5 code groups per minute, but only 6 percent could reach 15 (fig. 5.1).

Once learned and practiced routinely, however, sound receiving became as easy as listening to natural-language speech; one decoded automatically, going directly from sounds to word impressions. A woman who worked on Morse code receiving as part of the massive effort at Bletchley Park to decrypt German Enigma transmissions during World War II reported that after her intense experiences there, she heard Morse code everywhere—in traffic noise, bird songs, and other ambient sounds—with her mind automatically forming the words to which the sounds putatively corresponded.[4] Although no scientific data exist on the changes sound receiving made in neural functioning, we may reasonably infer that it brought about long-lasting changes in brain activation patterns, as this anecdote suggests.

If bodily capacities enabled the "miraculous" feat of sound receiving, bodily limitations often disrupted and garbled messages. David Kahn (1967) reports that "a telegraph company's records showed that fully half its errors stemmed from the loss of a dot in transmission, and another quarter by the insidious false spacing of signals" (839). (Kahn uses the conventional "dot" here, but telegraphers preferred "dit" rather than "dot" and "dah" rather than "dash," because the sounds were more distinctive and because the "dit dah" combination more closely resembled the alternating patterns of the telegraph sounder.) Kahn's point is illustrated in Charles Lewes's "Freaks of the Telegraph" (1881), in which he complained of the many ways in which telegrams could go wrong. He pointed out, for example, that in Morse code *bad* (dah dit dit dit [b] dit dah [a] dah dit dit [d]) differs from *dead* (dah

Figure 5.1 Army chart (1945) showing average times to learn sound receiving. Photograph by Nicholas Gessler.

dit dit [d] dit [e] dit dah [a] dah dit dit [d]) only by a space between the *d* and *e* in dead (i.e., _. . . . . _ _ . . versus _. . . . . _ _. .). This could lead to such confounding transformations as "Mother was bad but now recovered" into "Mother was dead but now recovered." Of course, in this case a telegraph operator (short of believing in zombies) would likely notice something was amiss and ask for confirmation of the message—or else attempt to correct it himself. Lewes gives many examples where the sender's compressed telegraphic language led the operator to "correct" the message, often with disastrous results. He cites the following example: "A lady, some short time since, telegraphed, 'Send them both thanks,' by which she meant, 'Thank you; send them both'—(the 'both' referred to two servants). The telegram reached its destination as 'Send them both back,' thus making sense as the official mind would understand it, but a complete perversion of the meaning of the writer" (470).

Given multiple possibilities for error, a strict discipline of inscription was instituted. Many code books contain exhortations to write each letter clearly, taking care to print them and leaving spaces between letters for better legibility. War Department technical manual TM11-459, *International Morse Code (Instructions)* (1945), for training telegraph operators goes further by showing in what order and direction letter strokes must be made (fig. 5.2). Disciplining the body in this way was one of many practices that made telegram writing an inscription technology enrolling human subjects into technocratic regimes characterized by specialized technical skills, large capital investments, monopolistic control of communication channels, and deferrals and interventions beyond the ken of the individual telegram writer and receiver.

The disciplinary measures extended to the construction of code as well, as bodily limitations forced adaptations in code book technologies. By the 1880s, many code books eliminated code words that differed by only one letter to reduce possible misreadings. In addition, after the widespread acceptance of five-letter artificial code words, beginning with *Bentley's Complete Phrase Code* in 1906, some code books contained the alarmingly named "mutilation tables" that allowed one to reverse-engineer errors and figure out what the correct letter must be. (Prior to that, code books employed "terminals" [endings] to help figure out the incorrect code group.) These measures also contributed to a sense that language was no longer strictly under one's control. Subject to a complex transmission chain and multiple encodings/decodings, telegraph language began to function as a nexus in which technological, social, and economic forces converged, interpenetrating the native expression of thought to create a discourse that always had multiple authors, even if originally written by a single person. Henry James's novella *In the Cage* (1898) brilliantly dramatizes how class distinctions, coding procedures, covert desires, and operator intervention intermingled to create a complex linguistic-social-technological practice (Menke 2008; Hayles 2005).

Although the electric telegraph, through its multilayered structure in which natural language cohabited with technical transmission signals and

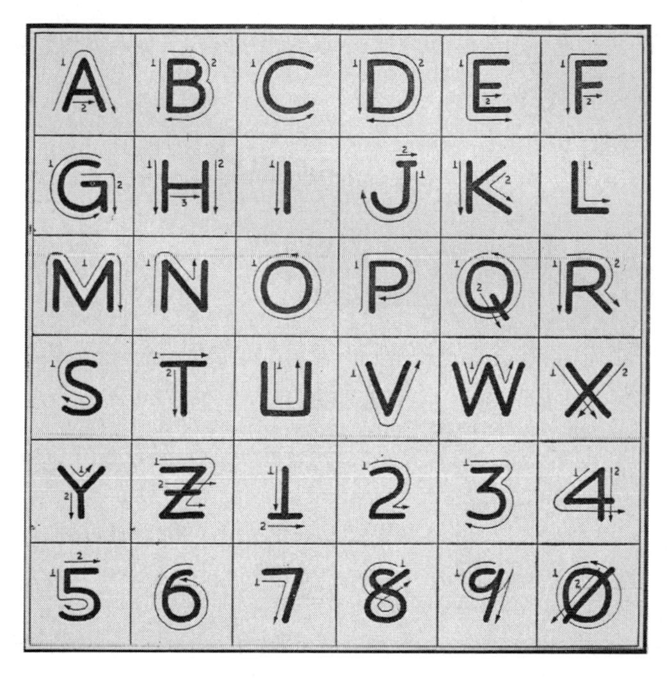

Figure 5.2 Army chart (1945) showing the directions in which letter strokes should be made. Photograph by Nicholas Gessler.

arbitrarily chosen code words, anticipated computer code, there is nevertheless a crucial difference between telegraph language and its evolutionary successors. The visible components of the acoustic electric telegraph, the key and sounder, required human intervention to operate. Unlike executable code of intelligent machines, telegraph code books did not give instructions to telegraph instruments. The goal articulated during the mid-twentieth-century Macy conferences of eliminating "the man in the middle" was never possible with the electric telegraph, making the technology intrinsically bound to the place of the human (Heims 1991; Hayles 1999).

Code and Language

Imagine you are sitting in your office preparing to write a telegram. Your employer, ever conscious of costs, has insisted that all office telegrams must be encoded. How would you be likely to proceed? Experience has taught you it is a waste of time to write out the entire telegram in plaintext and then encode it. Instead, you jot down a list of keywords corresponding to the gist of what you want to say and then reach for your copy of the company's official telegraph code book. Using the keyword list, you quickly look up the phrases listed under a given keyword, choosing the one that best suits your purpose. If the phrasing deviates slightly from what you might have intended, you are willing to accept it as "good enough," because it would take time and money to write out in plaintext a hypothetical alternative. To make the example more specific, suppose you work for a banking firm and have heard a rumor that a client asking for a loan is in financial trouble. Using the *Direct Service Guide Book and Telegraphic Cipher* (1939), you find under the keyword "Difficulties Rumored," the code word BUSYM, standing for "We have information here that this concern is in financial difficulties. An immediate investigation should be made. Send us the results." You had not intended to ask for an investigation, only confirmation or denial of the rumor. But seeing the phrase, you think perhaps it is not a bad idea to press for further information. You therefore write BUSYM and send off your telegram, confident it expresses your thoughts. This fictional scenario suggests that the code books, by using certain phrases and not others, not only disciplined language use but also subtly guided it along paths the compilers judged efficacious. In addition to inscribing messages likely to be sent, the code books reveal ways of thinking that tended to propagate through predetermined words and phrases, a phenomenon explored in more depth below.

In addition, code books instantiated a language regime that presented it as a multilayered system of plaintext and code equivalents. Friedrich Kittler (1992:25–35) has written about the transformation that took place as language moved from handwritten expression to typewriter, suggesting that the intimate connection of hand with pen encouraged an embodied sense of "voice" aligning the writer's subvocalization with a remembered maternal voice resonant with intimations of Mother Nature. With a typewriter, not only a different musculature and sense of touch but also a different appearance of letters was involved, achieving a mechanical regularity that seemed to some inhuman in its perfection.[5]

What then of code? Although telegrams typically began with a handwritten document (transitioning to typed documents toward the end of the nineteenth century), the encoding process enmeshed the user in a language system that presented in tabular form code words and/or numbers linked to natural-language phrases. The structure anticipates Ferdinand de Saussure's ([1916] 1983) theorizing the sign as comprising an arbitrarily connected signifier and signified, with an important difference. Rather than signifier and signified inhering in a single word, in the code books they are distributed between a code word (the signifier) and a corresponding word or phrase (the signified). This dispersion has significant theoretical consequences. Whereas Saussure's theory of *la langue* brought into view the important implication that meaning derives from paradigmatic differential relations between signs rather than direct connection of sign to referent, in the code books meaning derives from paradigmatic differences between phrases listed under a keyword, many of which are closely related (for example, a set of phrases under the keyword "sell" might read "will sell," "have already sold," "cannot sell," "offer to sell," etc.). As the example illustrates, encoded telegraphic communications were structured by keywords (typically listed in a keyword index), so that ideas flowed from a preset list of major concerns through to nuances provided by the different phrases under a keyword. Writing in this mode focused attention first on nouns representing typical categories, with actions contained in the phrase choices, or on verbs subsequently modified by adverbs, tenses, etc. Like the bullet list of contemporary PowerPoint presentations, communications filtered through code books were dominated by actors and actions, with time indicators, adjectives, adverbs, and other nuances and qualifiers relegated to secondary descriptors. In this sense, code books reflected the syntactical structure of English (as well as major European languages) and, in an American context, the autonomous agency associated with a capitalistic ideology of self-moving, self-actuated individualism.

Saussure's insight that the association between word and thing is arbitrary finds an antecedent of sorts in telegraph code books. Compilers generally devised code words that had no intrinsic connections with the plaintext phrases, a writing convention that facilitated the use of code books for secrecy. Assuming the general public did not have access to a given code book, a user could create a coded telegram that could not easily be decoded, for the arbitrary relation meant that the plaintext could not be guessed from the code word's ordinary meaning. Hence the injunction imprinted in many commercial code books to keep the books secure and safe from unauthorized users. Often individual code books were numbered, with accounts kept of which numbered book was assigned to which user. If a book turned up missing, the bureaucracy would know whom to blame (and conversely, the user to whom the book had been issued would know he would be held accountable for its security).

The issue of secrecy blended with more general concerns about privacy. For a public accustomed to distance communication in the form of sealed letters, the presumption of privacy often acquired moral force. Herbert Hoover's secretary of state Henry Stimson exemplified this attitude when, in

1929, he proposed eliminating the nascent US Government Cipher Bureau (charged with intercepting and decoding secret communications from other countries) on the grounds that "gentlemen do not read other people's mail" (qtd. in Kahn 1967:360). There are many anecdotes about the shock people received when they realized that telegrams, because of the technical transmission procedures involved, *required* that other people read personal telegrams. One such account has a society lady delivering her correspondence in a sealed envelope to the telegraph operator. When he ripped it open, she admonished him by declaring, "That is my *private* correspondence!" Robert Slater's *Telegraph Code, to Ensure Secrecy* (8th ed., 1929), puts the case this way in the introduction:

> On the 1st February, 1870, the telegraph system throughout the United Kingdom passes into the hands of the Government, who will work the lines by Post Office officials. In other words, those who have hitherto so judiciously and satisfactorily managed the delivery of our sealed letters will in future be entrusted also with the transmission and delivery of our open letters in the shape of telegraphic communications, which will thus be exposed not only to the gaze of public officials, but from the necessity of the case must be read by them. Now in large or small communities (particularly perhaps in the latter) there are always to be found prying spirits, curious as to the affairs of their neighbours, which they think they can manage so much better than the parties chiefly interested, and proverbially inclined to gossip. (1)

Slater's rhetoric must have been effective, for his code book went through nine editions, from 1870 to 1938, even though it failed to compress messages and had a limited vocabulary. His success indicates that the secrecy afforded by coded telegrams was widely seen as protecting not only privileged business dealings and confidential government affairs but also the privacy of ordinary citizens.

Notwithstanding the arbitrariness mentioned above, in some instances the compiler took advantage of a closer connection between code word and plaintext phrase. For example, the military foresaw situations in which it would be necessary to convey the information, in code, that the code book has been lost. In that case, one could easily run into the catch-22 of needing the code book to find the code indicating that the book was lost. The ingenious solution, emblazoned on the cover of *The "Colorado" Code* (1918) used by the US Army in World War I, is implied by the stern instruction

"MEMORIZE THIS CODE GROUP: 'DAM'—Code Lost." In other instances, the relation between code word and plaintext constitutes a Freudian peekaboo, teasing the reader with subconscious associations that may have induced the compiler to join code word and phrase in ways not altogether innocent: for example "Hastening" = "buy immediately" (Shepperson 1881:126), "Archly" = "Is very much annoyed" (Dodwell and Ager 1874:29), and "Hostile" = "Have you done anything" (*The Baltimore and Ohio Railway Company Telegraphic Cipher Code* 1927:56).

The printed spatial separation of code word and corresponding phrase translated in practice into a temporal separation. Rather than a single enunciative act such as Saussure analyzed, the encryption process took place as discrete moments of finding the keyword, choosing the phrase, and then writing down the appropriate code word. This temporal dispersion was spread further after encryption, as the telegram was translated into Morse code by the telegraph operator, sent as a signal over wires, decoded by the receiving operator, and finally decrypted by the recipient. Telegraphy had its own version of Derridean différance (deferral plus difference), for in its chain of translation and transmission, mistakes would happen at many points: losing one's place in the code book lineation and writing an incorrect code word; malforming letters so the operator misread one or more letters and mangled the meaning; failure to ask the operator to repeat the message for verification; operator error at the key, for example in dropping a dit which could change the meaning of an entire phrase; noise in the wires, causing the signal to be received in incorrect form; operator error at the sounder, misinterpreting the dits and dahs and transcribing them as an incorrect letter; and decryption errors by the recipient. What is true of all language from a deconstructive point of view is literalized and magnified in telegraphy through temporal and spatial dispersions and the consequent uncertainties and ambiguities they introduced. The dispersions also meant that the linguistic code system was necessarily enmeshed with technological and embodied practices.

As code books moved to algorithmic code constructions, the embodied practices of individuals directly involved with message transmissions began to spread through the population by changing assumptions about the relation of natural language to telegraph code. Not only telegraph operators but anyone who sent or received an encoded telegram (which included almost all businesspeople as well as many who used telegrams for social communications) participated in the linguistic practices that had bodily and cognitive effects.

As we have seen, language mediated through telegraphy was manifestly part of a technocratic regime that called into play different kinds of physico-cognitive practices than handwriting or typewriting alone. The first example (1906; fig. 5.3) demonstrating this regime shows a telegram encoded using a code book that correlates phrases with natural-language words, written by the receiving telegrapher and decoded in a different hand, presumably the recipient's. The second example (1900; fig. 5.4) shows an encoded telegram that someone has decoded by writing, in a different hand, the natural-language words and phrases above the code words. The next image (fig. 5.5) shows a telegram (1937) that uses made-up code words of five letters each. Unlike code books that used dictionary words for code and thus made it possible to spot misspellings, code books of the sort used in this case made correction through spelling impossible. Special charts known as "mutilation tables," discussed below, were necessary to check for errors.

Implicit in the evolution of code books as they moved from dictionary words to made-up code groups is a complex nexus in which embodied practices, economic imperatives, international politics, and technological con-

Form No. 3.

WESTERN UNION — Cable Message — WESTERN UNION

THE WESTERN UNION TELEGRAPH COMPANY.

INCORPORATED

ROBERT C. CLOWRY, President and General Manager.

THE LARGEST TELEGRAPH AND CABLE SYSTEM IN EXISTENCE. CABLE SERVICE TO ALL THE WORLD.

14,000 OFFICES AND 25,000 ADDITIONAL TELEGRAPH AND TELEPHONE CONNECTIONS IN NORTH AMERICA.

DIRECT AMERICAN CABLES NEW YORK TO GREAT BRITAIN.
CONNECTS also with ANGLO-AMERICAN and DIRECT U. S. ATLANTIC CABLES.
DIRECT COMMUNICATION WITH GERMANY AND FRANCE, CUBA, WEST INDIES, MEXICO and CENTRAL and SOUTH AMERICA.
WITH PACIFIC CABLES TO ALASKA, HONOLULU, AUSTRALIA, GUAM, THE PHILIPPINES, JAPAN, ETC.

Branch Offices in Principal Cities of Great Britain and the European Continent. All Foreign Telegraph Stations accept Messages to be sent "Via Western Union."

NUMBER	SENT BY	REC'D BY	No. OF WORDS	FROM
8NY	60	MF	6 350PM	Buenos Ayres

RECEIVED at Flushing, N.Y. Dec 4 1906 190

Lane Flushing N Y. 240

Nonentity anyhow.

Stop shipping & Pending further advice

Figures 5.3 Western Union telegram (1906) using natural-language code words with transcription below ("stop shipping"; "pending further advice"). Photograph courtesy of Fred Brandes.

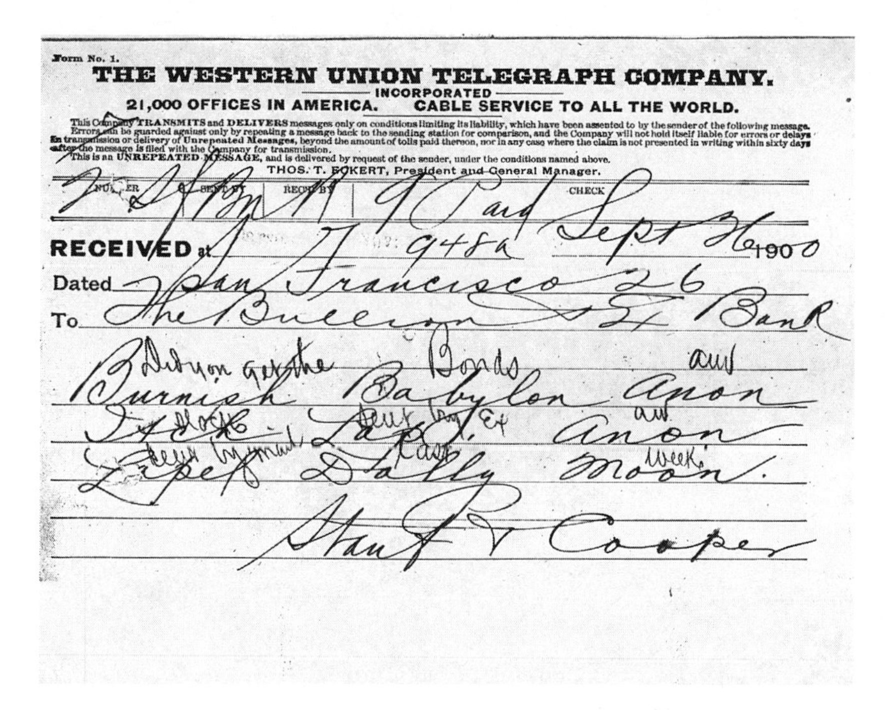
Form No. 1.

THE WESTERN UNION TELEGRAPH COMPANY.

INCORPORATED

21,000 OFFICES IN AMERICA. CABLE SERVICE TO ALL THE WORLD.

This Company TRANSMITS and DELIVERS messages only on conditions limiting its liability, which have been assented to by the sender of the following message. Errors can be guarded against only by repeating a message back to the sending station for comparison, and the Company will not hold itself liable for errors or delays in transmission or delivery of Unrepeated Messages, beyond the amount of tolls paid thereon, nor in any case where the claim is not presented in writing within sixty days after the message is filed with the Company for transmission.

This is an UNREPEATED MESSAGE, and is delivered by request of the sender, under the conditions named above.

THOS. T. ECKERT, President and General Manager.

NUMBER	SENT BY	REC'D BY		CHECK
			Paid	

RECEIVED at 948a Sept 26 1900

Dated San Francisco 26

To The Bullion & Ex Bank

Burnish Babylon Anon

Daily Moon.

Stauf & Cooper

Figure 5.4 Coded telegram (1900) with plaintext phrases decoded above the code words. Photograph courtesy of Fred Brandes.

straints and capabilities entwined. The earlier code books were compiled using human memory and associative links as part of the process of code construction. John W. Hartfield in the 1930s provides a glimpse into the construction of the plaintext phrases. In the following passage (surprisingly redundant for one who made his living from telegraphic compression), he tells his story through repetition, perhaps unconsciously performing the same kind of associative linking that he describes as his method: "I had a great mass of material accumulated from years past, different codes, and gleanings of suggestions made by different people and so forth. I took these and made notes of them on sheets of paper, writing phrases on sheets of paper. As I wrote phrases, other phrases suggested themselves and I interpolated them. I read the phrases and as I read them, other phrases suggested themselves, and I wrote those. Then I rewrote them into alphabetical sequence, and as I rewrote them into alphabetical sequences other phrases suggested themselves, and those I interpolated" (qtd. in Kahn 1967:847). Recalling Vannevar Bush's well-known assertion that the mind thinks associatively,

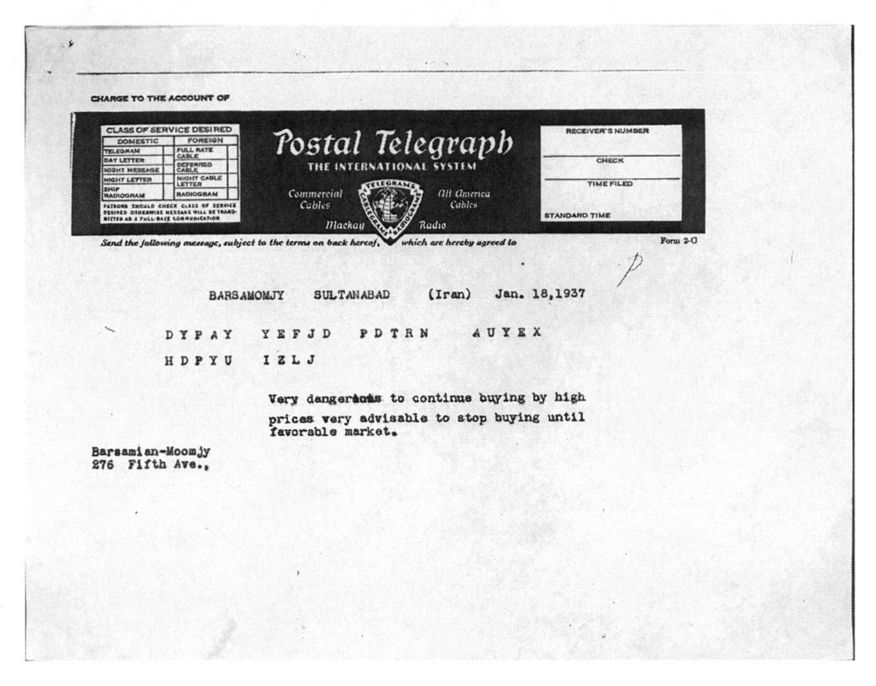
CHARGE TO THE ACCOUNT OF

Postal Telegraph
THE INTERNATIONAL SYSTEM

Commercial Cables — All America Cables — Mackay Radio

Send the following message, subject to the terms on back hereof, which are hereby agreed to

BARSAMOMJY SULTANABAD (Iran) Jan. 18,1937

DYPAY YEFJD PDTRN AUYEX
HDPYU IZLJ

Very dangerous to continue buying by high prices very advisable to stop buying until favorable market.

Barsamian-Moomjy
276 Fifth Ave.,

Figure 5.5 Telegram (1937) using artificial five-letter words with decoding below ("Very dangerous to continue buying . . ."). Note correction to the word "dangerous." Photograph courtesy of Fred Brandes.

we can see two processes at work in this account, ordering and associating. Ordering reveals what is not there, and associating provides phrases to fill perceived lacunae. The description indicates how closely tied code book compilations were to the lifeworld of the compiler.

They were also tied, of course, to technical experience. William J. Mitchell, compiler of the Acme Code, was admirably succinct when asked how he constructed his vocabulary: "By reading telegrams" (qtd. in Kahn 1967: 846). The American Railway Association, in constructing code for the use of members, followed a more bureaucratic procedure in the *Standard Cipher Code of the American Railway Association* (1906), soliciting suggestions from railway department officials and forming a committee to make selections ("Introductory," n.p.). Whether the compilations were achieved through association, technical experience, or bureaucratic procedures, memory and tacit knowledge were key to code construction from the 1840s to the 1880s.

As code books proliferated, their authors became increasingly aware of problems of the kind Lewes discussed, and they sought to take preventive measures. The limitations of embodied practices initiated further transformations in the technology. David Kahn enumerates some of these: "They employed experienced telegraphers to eliminate words telegraphically too similar. They deleted words that might make sense in the business in which the code was used" (840). *Liebèr's Five Letter American Telegraph Code* (1915), for example, advertised itself as offering "the greatest safety" for transatlantic cables, declaring, "The *selection* of Ciphers alone has entailed a labor of many months, including the assistance of skilled experts thoroughly versed in the probability of ordinary and even extraordinary mutilation in passing of messages from one circuit to another, to say nothing of the almost inevitable errors arising from frequent imperfect working of the ocean cables" (1). By the late 1880s, code book compilers boasted of including only words that differed from one another in spelling by at least two letters, thus reducing the probability of mistaking one code word for another. *Bentley's Second Phrase Code* (1920), for example, maintained "a minimum difference of two letters between two codewords" (iv; see also Kahn 1967:840). This principle, known as the "two letter differential," contributed further to limiting available code words, especially since authors also decided to strike foreign words from their code word vocabularies, reasoning that these were more difficult for telegraphers to encode correctly. Moreover, *Bentley's* also adhered to the principle that "if any two consecutive letters to any codeword are reversed the word so formed will not be found in the code" (iv). These prophylactic measures had a consequential outcome: because of these constraints, natural-language code words were no longer sufficient to match code words with all the desired phrases, so some authors began to invent artificial code words.

The first such words were formed by adding suffixes to existing words. Further inventions included adding a signifying syllable to an existing word; Kahn mentions that "*FI* meant *you* or *yours*, *TI* meant *it*, *MI* meant *me*, *I* or *mine*, etc." (1967:841). This opened a floodgate of innovation, which had the effect of moving code from natural language to algorithmically generated code groups. Typical were code condensers that allowed one to move from seven-figure code groups to five-letter groups (this reduction was possible because there are more distinct letters [26] than single-digit whole numbers [0–9]). *The Ideal Code Condenser* (n.d.), for example, comments that the reduction permitted by code condensers may be "frequently amounting to one-fourth [of the original telegram cost] or one-fifth of the former cost, or even less" (1). Many of the code books had number equivalents for the phrases, and code condensers were used to convert numbers to letters (an economic advantage, since each digit in a number was charged as if it were a word). The *ABC Code* (1901), for example, had a table of cipher codes (also known as merchant codes) that could be used to encipher the code words into number equivalents for greater secrecy. The last natural-word edition of the *ABC Code* (5th ed., 1901) had such a table along with instructions, whereas the sixth edition (1920) did away with this method, as artificial code words became predominant. Moreover, artificial letter codes could be constructed with a regular alternation between vowels and consonants, thus enabling some measure of self-correction if an error was made in coding.[6]

Behind many of these developments was not only the race of telegraph companies to maximize their profits (versus the opposed intentions of code book companies to save their customers money) but also the kinetics and capabilities of the human mind/body. As codes moved away from natural languages, the ability of operators to code and decode correctly was diminished. Presumably this was the reason that the International Telegraph Union (to which the major European countries were signatories, although not the United States) in its 1903 conference ruled that "On and after 1st July, 1904, any combination of letters not exceeding ten in number will be passed as a Code Word provided that it shall be pronounceable according to the usage of any of the languages to which code words have hitherto been limited, namely: English, French, German, Dutch, Italian, Spanish, Portuguese and Latin" (qtd. in *The Ideal Code Condenser* n.d.:1). That is, the syllables had to be pronounceable in one of the eight languages that the International Telegraph Union had ruled in 1879 were the only ones that counted as "natural" and hence exempt from the higher rate for enciphered telegrams. (The International Telegraph Convention distinguished between code telegrams, which could be decoded using the appropriate code book, and telegrams that used a cipher, defining them as "those containing series or groups of figures or letters having a secret meaning or words not to be found in a standard dictionary of the language" [qtd. in *ABC Telegraphic Code* 1901:vi]). From the telegraph companies' point of view, ciphers were difficult to handle and more sensitive to error, thus putatively justifying the extra charge; the spin-off result was that telegrams making secrecy a priority were charged at a higher rate. The list of languages considered "natural," noted above, gives the Western colonial powers privileged status, indicating one of the ways in which international politics became entangled with the construction of codes—along, of course, with regulations setting telegram tariffs, negotiations over rights to send messages on the lines of another company, and other capitalistic concerns.

Figure 5.6 Joseph Grassi's machine for producing pronounceable code groups. Photograph by Nicholas Gessler.

A rather quixotic response to the "pronounceable" rule was made by Joseph Grassi, who in 1923 applied for a patent on a cryptographic machine (fig. 5.6) that would automatically produce pronounceable code groups from plaintext. It did so by substituting vowels for vowels and consonants for consonants, which meant that there was no information compression, since it did not condense phrases into single words. It was also not very successful for secrecy; since the vowel-consonant pattern of the encoded text was the same as the plaintext, the code was relatively easy to break. The final blow to Grassi's ambitions came when, after he finally received a patent in 1928, the rule about pronouceability was rescinded in 1929. Nevertheless, the machine was beautifully engineered and has now become a prized collector's item, since it is quite rare (probably because the essentially useless machines were often discarded).

Grassi was not the only one to suffer from a lack of foresight; so did the International Telegraph Union when it formulated the rule that code words could be ten letters long. Even before the regulation was scheduled to take effect, *Whitelaw's Telegraph Cyphers: 400 Millions of Pronounceable Words* appeared in 1904 using five-letter artificial code groups (cited in Kahn 1967:843). This allowed users to combine two code groups to form a ten-letter word, which according to the new regulations was allowed to count as a single word, thus cutting in half the price of expensive transatlantic telegrams. The major code companies quickly followed suit in creating five-letter codes combinable to make a ten-letter word standing for two phrases spliced together.

The progression from natural language to artificial code groups, from code words drawn from the compiler's memory associations to codes algorithmically constructed, traces a path in which code that draws directly on the lifeworld of ordinary experience gives way to code calculated procedurally. To see the increasingly algorithmic nature of code after the 1880s, consider the mutilation tables at the back of *Bentley's Second Phrase Code* book ([1929] 1945) and reprinted here as figures 5.7, 5.8, and 5.9. Figure 5.7 gives the first two code letters of all possible code groups, figure 5.8 gives all possible combinations of the last two letters, and figure 5.9 lists all possible middle letters. Concatenating the three tables allows one to take an incorrect code word (known to be wrong because it does not appear in the book's list of code words) and reverse-engineer it to figure out a small set of possibilities (in the example *Bentley's* gives, there are a mere five code group possibilities), from which context would presumably allow the transcriber to choose the correct one. Such procedural algorithms are very far from the associative links that characterized early code construction.

As codes grew more procedural and decontextualized, the limitations of operator error became the bottleneck through which every message had to pass. The image in figure 5.10 of a 1941 telegram shows a correction to a code group; since *Bentley's* is specified as the code book used, it was no doubt corrected using *Bentley's* mutilation tables. This telegram is especially interesting because it was written after a decree issued during World War II that

AA	AB	AC	AD	AE	AF	AG	AH	AI	AJ	AK	AL	AM	AN	AO	AP	AR	AS	AT	AU	AV	AW	AX	AY	AZ
BE	BK	BV	BU	BH	BL	BW	BJ	BZ	BY	BR	BG	BI	BT	BD	BF	BM	BA	BX	BN	BB	BS	BC	BO	BP
CG	CX	CN	CJ	CW	CI	CP	CS	CK	CA	CC	CZ	CB	CO	CH	CM	CV	CL	CD	CY	CT	CF	CU	CE	CR
DV	DA	DW	DZ	DB	DU	DT	DK	DY	DR	DE	DN	DJ	DF	DI	DD	DH	DC	DL	DP	DS	DX	DG	DM	DO
EH	ER	EB	EN	EJ	EG	ES	EY	EP	EO	EM	EW	EZ	EX	EU	EL	EI	EE	EC	ET	EK	EA	EV	ED	EF
FR	FJ	FE	FL	FM	FX	FV	FI	FU	FZ	FY	FC	FD	FW	FF	FT	FO	FK	FS	FG	FH	FB	FA	FP	FN
GP	GU	GO	GA	GF	GK	GM	GL	GC	GG	GN	GR	GX	GH	GS	GB	GT	GZ	GJ	GE	GD	GI	GY	GW	GV
HY	HI	HR	HX	HO	HS	HE	HD	HL	HU	HZ	HA	HF	HV	HT	HW	HP	HJ	HB	HC	HM	HH	HK	HN	HG
IT	IG	IF	IR	IX	IY	ID	IC	IE	IV	IW	IO	IA	II	IK	IJ	IS	IN	IZ	IM	IL	IU	IP	IB	IH
JD	JP	JI	JV	JU	JE	JJ	JN	JW	JT	JF	JH	JG	JK	JC	JA	JL	JO	JR	JB	JZ	JY	JM	JX	JS
KJ	KM	KK	KT	KY	KW	KA	KO	KF	KD	KI	KS	KP	KC	KN	KG	KZ	KH	KV	KX	KR	KE	KB	KU	KL
LI	LO	LJ	LW	LZ	LV	LK	LP	LT	LF	LD	LB	LN	LA	LG	LC	LU	LM	LE	LS	LY	LR	LH	LL	LX
MU	MF	MZ	MB	MN	MH	MY	MT	MS	MX	ML	MJ	MW	MR	MV	ME	MG	MD	MM	MK	MP	MO	MI	MC	MA
NM	NY	NH	NG	NI	NC	NB	NZ	NN	NP	NO	NV	NU	NS	NL	NX	ND	NR	NA	NW	NJ	NK	NE	NF	NT
OX	OW	OL	OM	OC	OO	OU	OV	OH	OB	OS	OD	OE	OZ	OR	OY	OA	OT	OP	OI	OG	ON	OF	OK	OJ
PB	PE	PS	PP	PK	PN	PX	PR	PO	PM	PH	PT	PY	PL	PZ	PU	PJ	PV	PG	PF	PA	PC	PW	PI	PD
RO	RZ	RM	RC	RD	RA	RH	RU	RG	RN	RP	RE	RL	RB	RX	RS	RF	RY	RK	RV	RI	RJ	RR	RT	RW
SW	SC	ST	SY	SS	SZ	SF	SA	SR	SE	SV	SP	SK	SD	SJ	SI	SB	SG	SU	SO	SX	SL	SN	SH	SM
TZ	TD	TY	TS	TP	TB	TR	TF	TX	TL	TU	TK	TT	TE	TW	TV	TN	TI	TH	TA	TO	TM	TJ	TG	TC
UK	UH	UA	UF	UR	UT	UC	UM	UD	UI	UJ	UX	UO	UG	UP	UN	UY	UB	UW	UL	UE	UV	US	UZ	UU
VS	VV	VX	VO	VA	VP	VL	VE	VM	VH	VB	VF	VR	VU	VY	VZ	VK	VW	VN	VD	VC	VG	VT	VJ	VI
WL	WT	WU	WH	WG	WM	WZ	WW	WB	WS	WX	WI	WV	WY	WE	WR	WC	WF	WO	WJ	WN	WP	WD	WA	WK
XF	XN	XD	XE	XL	XR	XI	XG	XV	XW	XT	XM	XC	XJ	XA	XK	XX	XP	XY	XH	XU	XZ	XO	XS	XB
YN	YL	YP	YK	YT	YJ	YO	YX	YA	YC	YG	YY	YS	YM	YB	YH	YW	YU	YI	YR	YF	YD	YZ	YV	YE

SECTION I. FIRST TWO LETTERS.

Figure 5.7 *Bentley's* mutilation table (1945) showing all possible first two code letters.

K	CX	DA	ER	FJ	GU	HI	IG	JP	KM	LO	MF	NY	OW	PE	RZ	SC	TD	UH	VV	WT	XN	YL	ZS
V	CN	DW	EB	FE	GO	HR	IF	JI	KK	LJ	MZ	NH	OL	PS	RM	ST	TY	UA	VX	WU	XD	YP	ZG
U	CJ	DZ	EN	FL	GA	HX	IR	JV	KT	LW	MB	NG	OM	PP	RC	SY	TS	UF	VO	WH	XE	YK	ZI
H	CW	DB	EJ	FM	GF	HO	IX	JU	KY	LZ	MN	NI	OC	PK	RD	SS	TP	UR	VA	WG	XL	YT	ZV
L	CI	DU	EG	FX	GK	HS	IY	JE	KW	LV	MH	NC	OO	PN	RA	SZ	TB	UT	VP	WM	XR	YJ	ZD
W	CP	DT	ES	FV	GM	HE	ID	JJ	KA	LK	MY	NB	OU	PX	RH	SF	TR	UC	VL	WZ	XI	YO	ZN
J	CS	DK	EY	FI	GL	HD	IC	JN	KO	LP	MT	NZ	OV	PR	RU	SA	TF	UM	VE	WW	XG	YX	ZB
Z	CK	DY	EP	FU	GC	HL	IE	JW	KF	LT	MS	NN	OH	PO	RG	SR	TX	UD	VM	WB	XV	YA	ZJ
Y	CA	DR	EO	FZ	GG	HU	IV	JT	KD	LF	MX	NP	OB	PM	RN	SE	TL	UI	VH	WS	XW	YC	ZK
R	CC	DE	EM	FY	GN	HZ	IW	JF	KI	LD	ML	NO	OS	PH	RP	SV	TU	UJ	VB	WX	XT	YG	ZA
G	CZ	DN	EW	FC	GR	HA	IO	JH	KS	LB	MJ	NV	OD	PT	RE	SP	TK	UX	VF	WI	XM	YY	ZU
I	CB	DJ	EZ	FD	GX	HF	IA	JG	KP	LN	MW	NU	OE	PY	RL	SK	TT	UO	VR	WV	XC	YS	ZH
T	CO	DF	EX	FW	GH	HV	II	JK	KC	LA	MR	NS	OZ	PL	RB	SD	TE	UG	VU	WY	XJ	YM	ZP
D	CH	DI	EU	FF	GS	HT	IK	JC	KN	LG	MV	NL	OR	PZ	RX	SJ	TW	UP	VY	WE	XA	YB	ZM
F	CM	DD	EL	FT	GB	HW	IJ	JA	KG	LC	ME	NX	OY	PU	RS	SI	TV	UN	VZ	WR	XK	YH	ZO
M	CV	DH	EI	FO	GT	HP	IS	JL	KZ	LU	MG	ND	OA	PJ	RF	SB	TN	UY	VK	WC	XX	YW	ZE
A	CL	DC	EE	FK	GZ	HJ	IN	JO	KH	LM	MD	NR	OT	PV	RY	SG	TI	UB	VW	WF	XP	YU	ZX
X	CD	DL	EC	FS	GJ	HB	IZ	JR	KV	LE	MM	NA	OP	PG	RK	SU	TH	UW	VN	WO	XY	YI	ZF
N	CY	DP	ET	FG	GE	HC	IM	JB	KX	LS	MK	NW	OI	PF	RV	SO	TA	UL	VD	WJ	XH	YR	ZZ
B	CT	DS	EK	FH	GD	HM	IL	JZ	KR	LY	MP	NJ	OG	PA	RI	SX	TO	UE	VC	WN	XU	YF	ZW
S	CF	DX	EA	FB	GI	HH	IU	JY	KE	LR	MO	NK	ON	PC	RJ	SL	TM	UV	VG	WP	XZ	YD	ZT
C	CU	DG	EV	FA	GY	HK	IP	JM	KB	LH	MI	NE	OF	PW	RR	SN	TJ	US	VT	WD	XO	YZ	ZL
O	CE	DM	ED	FP	GW	HN	IB	JX	KU	LL	MC	NF	OK	PI	RT	SH	TG	UZ	VJ	WA	XS	YV	ZR
P	CR	DO	EF	FN	GV	HG	IH	JS	KL	LX	MA	NT	OJ	PD	RW	SM	TC	UU	VI	WK	XB	YE	ZY

Figure 5.8 *Bentley's* mutilation table (1945) showing all possible last two code letters.

T	X	D	L	C	S	J	B	Z	R	V	E	M	A	P	G	K	U	H	W	N	O	Y	I	F
V	B	T	S	K	H	D	M	L	Z	R	Y	P	J	G	A	I	X	O	E	C	N	U	F	W
A	E	G	V	H	R	P	Y	T	D	J	I	U	M	X	B	O	W	Z	K	S	L	F	N	C
F	L	I	U	G	X	K	S	Y	E	W	V	H	C	O	N	A	Z	B	T	P	M	R	J	D
R	M	V	H	I	O	T	P	S	L	Z	U	G	D	A	J	F	B	N	Y	K	C	X	W	E
Y	O	E	M	D	P	W	N	B	X	U	L	C	F	K	I	T	H	G	Z	J	A	S	V	R
M	I	B	J	Z	D	X	F	A	G	P	N	W	U	E	Y	L	K	T	O	R	V	C	S	H
L	G	Z	N	W	C	R	A	O	H	S	B	J	V	D	T	E	P	K	X	F	I	M	Y	U
C	V	N	W	B	E	O	R	F	I	K	J	Z	H	L	S	M	T	Y	A	X	U	D	P	G
H	J	S	K	Y	I	L	D	C	N	O	P	T	Z	V	R	U	A	F	M	E	W	G	X	B
Z	P	R	O	F	N	V	G	H	S	L	X	A	T	J	D	W	M	C	U	I	K	B	E	Y
X	C	U	G	V	A	Y	K	P	M	B	H	I	E	F	W	R	N	J	S	T	D	O	Z	L
O	D	H	I	U	F	S	T	K	C	N	G	V	L	R	Z	X	J	W	P	Y	E	A	B	M
U	N	Y	P	T	G	E	C	M	B	X	S	K	W	I	F	V	O	A	L	D	J	H	R	Z
K	R	C	E	M	Y	N	Z	W	F	I	D	L	O	S	H	P	V	U	J	B	X	T	G	A
W	S	F	X	A	B	I	H	U	Y	E	R	O	K	N	C	J	L	M	V	G	P	Z	D	T
S	A	L	C	E	K	Z	J	N	O	H	M	D	R	T	V	Y	G	I	B	W	F	P	U	X
J	Y	A	R	O	Z	G	U	V	T	D	F	X	P	B	M	N	E	L	I	H	S	W	C	K
B	K	X	A	R	J	U	I	G	P	M	O	F	Y	W	E	Z	C	D	H	V	T	N	L	S
E	H	W	B	J	M	F	O	X	U	Y	Z	N	I	C	K	D	S	P	R	A	G	L	T	V
D	U	J	Z	N	L	A	X	R	V	T	W	B	G	M	P	C	Y	S	F	O	H	E	K	I
G	W	P	T	S	V	M	E	D	J	A	K	Y	B	U	X	H	F	R	C	L	Z	I	O	N
P	F	M	D	L	T	B	W	J	A	G	C	E	X	Y	U	S	I	V	N	Z	R	K	H	O
N	T	O	F	X	W	H	V	I	K	C	A	R	S	Z	L	B	D	E	G	U	Y	J	M	P
I	Z	K	Y	P	U	C	L	E	W	F	T	S	N	H	O	G	R	X	D	M	B	V	A	J

Figure 5.9 *Bentley's* mutilation table (1945) showing all possible middle letters.

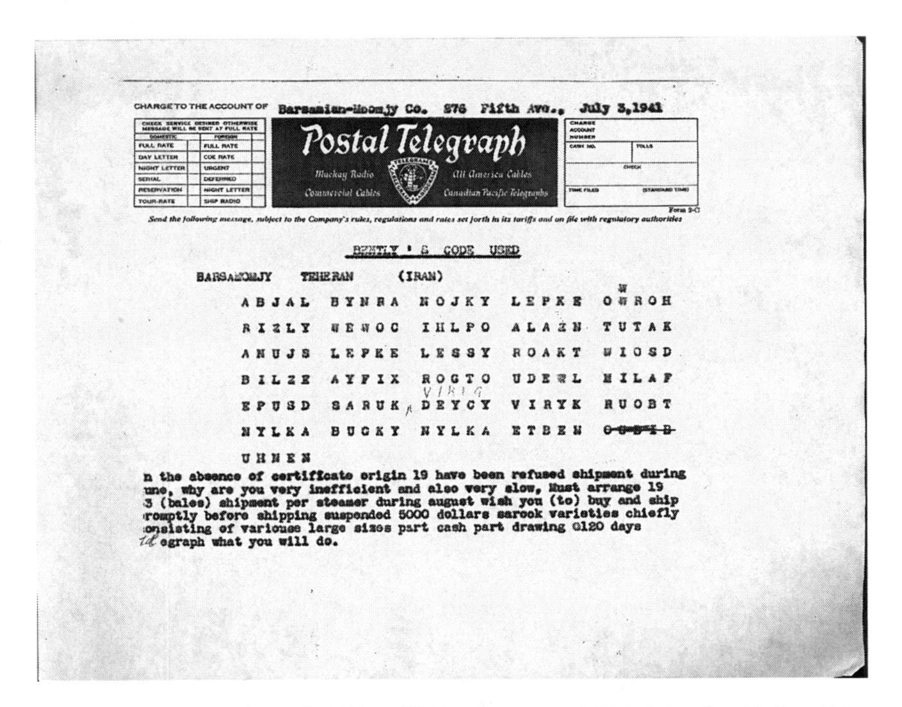

CHARGE TO THE ACCOUNT OF Barsamian-Moomjy Co. 276 Fifth Ave., July 3,1941

Postal Telegraph

Mackay Radio · All America Cables · Commercial Cables · Canadian Pacific Telegraphs

Send the following message, subject to the Company's rules, regulations and rates set forth in its tariffs and on file with regulatory authorities

BENTLY'S CODE USED

BARSAMOOMJY TEHERAN (IRAN)

ABJAL BYNRA NOJKY LEPKE OWROH

RIZLY WEWOC IHLPO ALAZN TUTAK

ANUJS LEPKE LESSY ROAKT WIOSD

BILZE AYFIX ROGTO UDEWL NILAF

EPUSD SARUK DEYCY VIRYK RUOBT

NYLKA BUCKY NYLKA ETBEN

UHNEN

n the absence of certificate origin 19 have been refused shipment during une, why are you very inefficient and also very slow, Must arrange 19 3 (bales) shipment per steamer during august wish you (to) buy and ship romptly before shipping suspended 5000 dollars sarook varieties chiefly onsisting of variouse large sizes part cash part drawing @120 days tel egraph what you will do.

Figure 5.10 Telegram (1941) with five-letter artificial code groups and plaintext decoding at bottom. Note that *Bentley's* (misspelled) is specified as the code book used. Photograph courtesy of Fred Brandes.

all encoded telegrams had to specify the code book used, a measure to make sure that telegrams could be monitored. (The recipient cannot have read the telegram with much pleasure, since the decoding typed at the bottom asks him "why are you very inefficient and also very slow.")

The dynamic between being bound by human limitations and liberated from them through technological innovations is nicely captured by the Vibroplex telegraph key, invented by Horace G. Martin in 1905. The device, a semiautomatic electromechanical key, repeats dits for as long as the operator presses the key to the right, thus reducing the hand motions required and consequently speeding up transmission times. The dahs, however, are sent individually by pressing the key to the left. Other innovations included telegraphs that directly printed messages without the necessity of either Morse code or sound receiving. Such was the Hughes telegraph, invented by David Hughes in 1855. The Hughes telegraph featured a piano-like keyboard marked with twenty-eight letters and punctuation marks. In a sense, piano technology, particularly tuning forks and rhythmic hammer strikes, provided the underlying analogy for the machine's operation. When the operator hit a key, a hammer struck a whirling wheel inscribed with fifty-six characters (letters, numbers, and punctuation marks), which in turn sent out an electrical pulse. In the sending instrument, a vibrating spring established the rate at which its wheel revolved; a spring tuned to the same rate of vibrations controlled the wheel movement of the receiving instrument. The two instruments were synchronized through the transmission of a dash; each time a dash (the beginning setting for the wheel) was transmitted, the receiving wheel was synchronized to the same position. The time interval between the dash and the hammer strike indicated the letter to be struck by the receiving instrument. The wheel rotated at about 130 revolutions per minute; despite this speed, the synchronization of the two spatially separated machines was found to vary by no more than one-twentieth of a second. The synchronization of clock times necessitated by interregional telegraph networks (Carey 1989) and later by the Teletype is here instantiated on a microscale. George B. Prescott, writing in 1860 about the Hughes telegraph, enthusiastically called it a "very ingenious contrivance" and "incomparable instrument" (n.p., chapter 9). Nevertheless, it still required considerable skill to operate, for the operator had to move in rhythmic synchronization with the speed of the wheel, which is perhaps one reason that it lost market share to the simpler and mechanically less complex Morse instrument. It also cost more, going for $130 in 1860.

The move toward the automation of message sending and receiving culminated in the teleprinter, a machine that constructed the code itself. The teleprinter was invented (over the period 1902–13) by the Canadian Frederick G. Creed, a telegrapher interested in finding a way to do away with the necessity of sending and receiving Morse code. With a keyboard reminiscent of a typewriter, the machine converted each keystroke to a series of electrical impulses, sending them over telegraph wires to a comparable machine that interpreted the impulses to reproduce the original keystrokes. The receiving instrument printed onto a gummed tape, which was then pasted onto a telegram form. As we have seen, sound receiving was a difficult skill to learn, so a trade-off was beginning to take shape: the skills required of the operator decreased as the machines grew more complex, while the skills required to produce and maintain the machines increased accordingly.

Although Creed's machine failed to take the world by storm, the American Charles L. Krumm developed an improved version, which, under the auspices of the American Telephone and Telegraph Company, was widely marketed in the 1920s under its trade name, the Teletype. The Teletype used the code devised by Émile Baudot for the multiplex telegraph (a telegraph that could send four signals at once, two in one direction down the wire, two in the other direction). Other teleprinters were devised that used a different keyboard code, the American Standard Code for Information Interchange, or ASCII. The five-bit Baudot codes gradually gave way to the seven-bit ASCII codes, still used today in formats integral to most computers and software systems (including "text" documents with the .txt extension). The telegraph is thus linked in a direct line of descent to the invention of computer code.

Through cryptographic practices, the telegraph is also linked to the invention of the digital computer, anticipated in the decryption machine called the Bombe, designed by a group of brilliant Polish mathematicians and then passed on to the British when the Nazi invasion of Poland loomed. That work gave Alan Turing and his colleagues at Bletchley Park a head start in World War II in their attempt to break German Enigma codes, transmitted via radio in Morse code. This cryptographic feat, in addition to shortening the war, provided the inspiration for the digital computer. With automatic spelling correction, electric keyboards, and other functionalities digital computers make possible, the importance of embodied capacities, although never absent, became less crucial to message transmission. Following the same kind of trajectory as the transition from sound receiving to Teletyping, fewer sending and receiving skills were located in humans, and more were located in the machines.

Information and Cultural Imaginaries of the Body

Before the balance tipped toward machine cognition, a zone of indeterminacy was created between bodies and messages. The electric telegraph was increasingly understood as analogous to the body in that it operated on electric signals dispersed through the "body" of the nation. Laura Otis (2001), in her history of this connection, remarks that "metaphors do not 'express' scientists' ideas; they *are* the ideas. . . . To physiologists, the telegraph and associated studies in electromagnetism suggested the mechanisms by which the body transmitted information. To engineers designing telegraph networks, organic structures suggested ways to arrange centralized systems. More significantly, they motivated societies to establish more connections in the hope of achieving a near-organic unity. . . . In the nineteenth century, the real 'language of communication' was metaphor itself" (48). The seminal importance of metaphoric connections between telegraphy and neuronal science suggests another way in which the technogenetic spiral works: models of the nervous system provide clues for technological innovation, and technological innovation encourages the adoptions of models that work like the technology. We saw in the previous chapter the uncanny resemblance between contemporary neuronal models emphasizing plasticity and economic business models that emphasize flexibility. As Catherine Malabou observes, in such cases it is important to look for gaps or ruptures that prevent the smooth alignment of such metaphoric understandings. In the case of telegraphy, one of the gaps that allowed constructive intervention was the "mysterious" nature of electricity. It was known in the mid-nineteenth century that nerves carried electrical impulses, and of course it was known that the telegraph also operated on electricity, which, however, did not prevent the telegraph from seeming nearly miraculous for many ordinary people. It is not surprising that in 1859 John Hollingshead should remark of the telegraph that "its working is secret and bewildering to the average mind" (Hollingshead 1860:235; qtd. in Menke 2008:163).

No doubt part of this bewilderment sprang from the new configurations of bodies and messages that the telegraph made possible. As previously noted, James Carey (1989) has argued that the telegraph "permitted for the first time the effective separation of communication from transportation" (203).[7] Prior to the telegraph, messages traveled only as fast as the medium of transportation that carried them: ship, train, horse, pedestrian. Carey outlines the many effects of the separation, among which were the establishment of regional and national market pricing, the dematerialization of goods as receipts (or information) substituted for physical objects, and a spiritualization of the technology that equated the mysterious "electric fluid" with the soul and the visible apparatus with the body. Laura Otis, translating the German physiologist Emil du Bois-Reymond writing in 1887, explicates how elaborate this analogy could be. Du Bois-Reymond wrote: "Just as . . . the electric telegraph . . . is in communication with the outermost borders of the monarchy through its gigantic web of copper wire, just so the soul in its office, the brain, endlessly receives dispatches from the outermost limits of its empire through its telegraph wires, the nerves" (qtd. in Otis 2001:49). The metaphor here is not merely fanciful, for indeed, the administration of colonies by remote metropolitan centers was transformed by telegraphy, allowing micromanagement of affairs that was impossible when slower means of communication were necessary.

In the new regime the telegraph established, a zone of indeterminacy developed in which bodies seemed to take on some of the attributes of dematerialized information, and information seemed to take on the physicality of bodies. Nelson E. Ross, in "How to Write a Telegram Properly" (1927), recounts an anecdote in which a "countryman" wanted to send boots to his son: "He brought the boots to the telegraph office and asked that they be sent by wire. He had heard of money and flowers being sent by telegraph, so why not boots? [The operator] told the father to tie the boots together and toss them over the telegraph wire. . . . [The countryman] remained until nightfall, watching to see the boots start on their long journey . . . [but] during the night, some one stole the boots. When the old man returned in the morning, he said: 'Well, I guess the boy has the boots by now'" (12). The old man's mistake was merely in thinking that he could supply the boots, for in the same booklet, Ross writes about the "telegraphic shopping service," which "permits of the purchase by telegraph of any standardized article from a locomotive to a paper or pins. The person wishing to make the purchase has merely to call at the telegraph office, specify the article he wishes to have bought, and pay the cost, plus a small charge for the service. Directions will then be telegraphed to the point at which the purchase is to be made, and an employee will buy the article desired" (13). The purchase is then either delivered in the same manner as a telegram or, if the recipient lives elsewhere, forwarded by parcel post. Thus bodies and information interpenetrate, with information becoming a body upon completion of the transaction.

Bodies and information entwined in other ways as well. It was no accident that telegraphy and railroads were closely aligned in the period from 1880 on, when monopoly capital seized control of regional telegraph companies and consolidated them. Railroad rights-of-way provided convenient places for the positioning of telegraph lines and also assured access for areas that otherwise might be difficult to service. In addition, railroad money was able to underwrite the large capital costs necessary to make a telegraph company a going concern, including the cost of stringing telegraph lines, establishing offices, and paying personnel. Railroads and telegraph messages were further entwined through the unprecedented mobilities that bodies and messages were achieving. Trains carried passengers and mail (i.e., bodies and information), but they also intersected with telegraph messages. Ross (1927) writes about the practice of addressing a telegram to a train passenger, cautioning the reader to include the railroad's name, train number, and place where the message should be delivered, as well as usual information such as the addressee's name (25–26). Railroad depots often included telegraph stations, and in smaller places, the station operator sometimes doubled as the telegraph operator. When a depot received a telegram addressed to a passenger on an incoming train, the operator (or his surrogate) would jump on board when the train stopped and deliver the message to the appropriate person.

At the same time, the practice of addressing telegrams "properly," as Ross puts it, reveals the globalization of the local that Carey (1989) discusses in terms of establishing national and regional market prices and standardizing time. The increasing national recognition that large companies received as a result of faster and more widespread communication networks made them nationally and not merely locally known sites. Ross comments, for example, that "if you are telegraphing a business firm of national prominence, a motor manufacturing company in a large automobile center, a famous bank, or for that matter any bank, or any manufacturing concern of widespread repute, a street address is not needed. Imagine, for example, the absurdity of giving a street address in a telegram to the General Electric Company in Schenectady" (13). Local knowledge thus mediated between the increasing spatial scope of communication that telegraphy made possible and the materialities involved in physical locations.

The racial implications of the dynamic between information and the cultural imaginaries of bodies are explored by Paul Gilmore (2002) in relation to Stephen Foster's well-known song "O Susanna," the first verse of which includes such whimsical lines as

It rain'd all night the day I left,
The weather it was dry,
The sun so hot I froze to death
Susanna, don't you cry.[8]

Gilmore points out that the second verse is much less known than the first. The rarely heard second verse is also much less charming:

I jumped aboard de telegraph
And trabbeled down de ribber,
De 'lectric fluid magnified,
And killed five hundred nigger.

Focusing on the grotesque and violent nature of the imagery, Gilmore argues, "Because electricity was understood as both a physical and spiritual force, the telegraph was read both as separating thought from body and thus making the body archaic, and as rematerializing thought in the form of electricity, thus raising the possibility of a new kind of body. . . . The telegraph's technological reconfiguration of the mind/body dualism gave rise to a number of competing but interrelated, racially inflected readings" (806). Gilmore's argument is compelling, especially since the telegraph was used by imperialist powers to coordinate and control distant colonies on a daily basis, a feat that would have been impossible had messages traveled more slowly by ship or train. In this sense too, the telegraph was deeply implicated in racist practices.

His reading of Foster's song, however, is at best incomplete.[9] He does not cite the remainder of the second verse, which suggests a very different interpretation:

De bullgine bust, de horse run off
I really thought I'd die,
I shut my eyes to hold my breath,
Susanna, don't you cry!

"Bullgine" was shipboard slang for a ship's engine, usually used in a derogatory sense. Ships were often pulled along shallow sections of the Mississippi and other rivers by horses. Electric fluid, although associated with the telegraph, was also commonly used to describe lightning, which became an increasing hazard as shipbuilding moved from wooden to iron construction, with the result that lightning strikes became common occurrences on riverboats, killing in several instances hundreds of people. In *Old Times on the Upper Mississippi*, George Byron Merrick (1909) reports that *Telegraph* was such a common name for riverboats that there was "great confusion of any one attempting to localize a disaster that had happened to one of that name in the past" (230). These facts suggest a more straightforward reading of the enigmatic second verse. The speaker jumped on a riverboat named *Telegraph*, which was struck by lightning, frightening the horse pulling the boat and killing "five hundred nigger."

This interpretation does not, of course, negate the racial violence depicted in the verse, nor does it explain why the speaker uses the rather obscure phrase "de 'lectric fluid magnified" rather than simply calling it a lightning strike. Rather, the gruesome imagery, the nonsensical nature of the first stanza, and the paradoxical line in the second ("I shut my eyes to hold my breath") suggest that both interpretations are in play, the commonsensical and the mysterious. There is an oscillation between reading "telegraph" as a ship (in which case there was nothing magical about the disaster) and "telegraph" as a communication technology in which bodies could be transported along telegraph lines as if they were dematerialized messages, albeit with fatal consequences if the " 'lectric fluid" happened to "magnify." The whimsically paradoxical nature of the lyrics now can be seen in another light. They insist on the necessity of holding two incompatible thoughts together in mind at once, as if anticipating the oscillation between commonsensical understanding of telegraphy as an everyday technology and as a mysterious reconfiguration of human bodies and technics.

Telegraph code books embody a similar kind of ambiguity. On the one hand, they were used in straightforward business practices to save money. On the other hand, through their information compression techniques, their separation of natural-language phrases from code words, and the increasingly algorithmic nature of code construction, they point the way toward a dematerialized view of information that would, a century beyond their heyday, find expression in the idea that human minds *already* exist as dematerialized information patterns and so can be uploaded to a computer without significant loss of identity (Moravec 1990, 2000). Norbert Wiener, writing at the dawn of the computer age in *The Human Use of Human Beings* (1954), carried this dynamic to its logical extreme when he speculated whether it would be possible to telegraph a human being (103). A number of writers have pointed to the risky nature of such a dematerialized view of human bodies, ranging from Francis Fukuyama in *The End of History and the Last*

Man (2006) to Greg Egan in *Permutation City* (1995). In historical context, they are recapitulating in other keys the explosive (and racially inflected) disaster conjured in Stephen Foster's minstrel song.

Messages and the Technological Unconscious

There are insufficient data to determine which code phrases were actually used in telegrams (as distinct from being published in the code books). Nevertheless, we may presume there was some correlation, especially since the compilers solicited feedback from those who purchased their books and constantly sought to make them correspond to prevailing usage. Even if the correlation is weak, however, the phrases still indicate attitudes, dispositions, and assumptions held by their authors. In this sense, they constitute further evidence about the entwinement of cultural assumptions with telegraphic technologies in technogenetic cycles that link code constructions with embodied practices and cultural imaginaries.

Particularly relevant in this regard are plaintext phrases that reveal how the telegraph was changing expectations and assumptions. The necessity of a speedy response, for example, is implicit in the plaintext phrases listed under the keyword "Strikes" in the *General Cipher Code* of the Railway Express Agency (1930): "Sticta—Was arrest made"; "Sigmaria—Ammunition and arms wanted"; "Stilbite—Armed with gatling guns"; "Stile—Armed with pistols"; "Stillhunt—Armed with rifles"; "Stillicide—Arms are wanted"; "Siffiform—Arrest party quickly and advise." Notice the emphasis on armed responses and punitive measures, with no hint that the strikers might have valid grievances. By comparison, *Liebèr's Telegraphic Code* (1915) gives the following phrases under "Strikes": "Kwyfa—A strike has been ordered (by the ——)"; "Kwyhf—Advise accepting the strikers' conditions"; "Kwyjl—At a meeting of the employers it was determined to resist the strike"; "Kwyko—At a meeting of the men it was determined to continue the strike"; "Kwyny—Do not see how we can escape giving in to the strikers." The last phrase, suggesting that negotiation with the strikers might be necessary, reflects a more balanced view in which different interests are contending, without presuming at the outset that the "right" side will necessarily be that of the railway authorities. The differences between the *Liebèr's* code book, intended for a general audience, and the more narrowly focused *General Cipher Code*, intended for use by railway officials and administrators, reveal contrasting sets of assumptions about which messages might be necessary.

Significantly, the Railway Express Agency regarded its code book as private information not to be shared with unauthorized users. The book I perused (NG) contains a number stamped on the cover to ensure accountability (a common practice with commercial code books, as mentioned earlier), along with the stern warning "This book is for the use of the person to whom it was issued and must not be left accessible to the view or examination of persons not authorized to use it," followed by the admonition (in capital letters) "KEEP UNDER LOCK AND KEY." The injunction suggests that the code book should be understood as proprietary information that, if revealed to outsiders, might have negative effects for the company, as indeed might be the case for codes anticipating armed responses to strikers. Code books stressing secrecy had the effect of dividing the population according to those who had access to information communicated over a common carrier and those who did not, anticipating the access codes and passwords of the contemporary web.

Other code books remind us of the historical circumstances under which disciplinary functions were carried out. *The Peace Officers' Telegraph Code Book* (Van Alstine 1911; FB), for example, has three pages showing different shapes of hats, each identified with a code group. In a time when it was difficult to send images, the hat pages substituted for merely verbal descriptions and provided more precise information than would have been possible with words alone. Here we see the realization emerging that images and messages could be correlated through intervening code, another anticipation of contemporary message protocols. In other cases, words had to suffice. There are several entries for identifying horses, including "a young horse," "a valuable horse," "a blind horse," "a horse with shoe missing," "a lame horse." Under "Deformities" are listed, among other entries, "club foot," "both left arm and leg off," "hump back," "round shoulder," "gunshot wound," "false teeth," "earrings in ears." Contemporary attitudes are conveyed with such entries as "Fiend," which includes "JWC—Automobile fiend," "JWD—Baseball fiend," "JWE—Cocaine and morphine fiend," "JWF—Cigarette fiend," "JWI—Cigar fiend," "JXS—Dope fiend," and my favorite, "JXO—Candy fiend." The specificity of these entries suggests that law officers were beginning to understand that identifying characteristics across a broad range of traits could be quickly communicated, with a corresponding emphasis on accurate descriptions and telling details.

The perils of shipping are vividly inscribed in codes devoted to that trade. The presumption of speed is apparent in *Watkins Universal Shipping Code* (Stevens 1884), which lists codes for "Involucred—SHIP a total loss" (804); "Irlandais—AFTER compartments are full of water" (805); "Ischnote—NO hope of getting her off. She is a total loss" (905); "JALOUX—THE sea is breaking completely over the ship" (807); "Jobot—SHE struck bottom in the river" (807); and the apparently paradoxical "Kabala—SHE had to be scuttled to save her" (811). More general code books had similar dire messages listed in their codes. *The ABC Universal Commercial Electric Telegraphic Code: Specially Adapted to the Use of Financiers, Merchants, Shipowners, Underwriters, Engineers, Brokers, Agents, &c., Suitable For Everyone* (5th ed., 1901) listed the codes "15938 Detallases—The Captain is drunk"; "15940 Deturceim—The Captain is continually drunk"; and the *Moby Dick*–ish "07975 Cuixilho—Captain is insane" (130). *The Ship Owner's Telegraphic Code* (Scott 1882), one of several shipping-oriented books published by E. B. Scott, has the usual catastrophic messages but also includes the intriguing "Hazard—crew saved by rocket apparatus" (162). These examples show that the presumption of fast communication was changing how people thought about intervening in rapidly changing situations.

Military code books are particularly chilling. The first widespread American military conflict in which the telegraph played an important role was the Civil War. David Home Bates, author of *Lincoln in the Telegraph Office* (1907), identifies himself on the title page as "Master of the War Department Telegraph Office, and Cipher-Operator, 1861–1866," indicating by the latter phrase the importance of crypto-encipherment and cryptanalysis for the war effort. Telegraph lines were tapped to intercept enemy dispatches by both Northern and Southern operatives, so the race was on to develop secure codes that the enemy could not break while striving to do just that to the opposing side. The telegraph, for the first time in a large-scale American conflict, allowed minute-by-minute reporting and communication between the commander in chief and his generals. Bates has many anecdotes about Lincoln lingering in the telegraph office, such as the following: "There were many times when Lincoln remained in the telegraph office till late at night, and occasionally all night long. One of these occasions was during Pope's short but disastrous campaign, ending in the second battle of Bull Run. Lincoln came to the War Department office several times on August 26, the first of those strenuous, anxious days, and after supper he came again, prepared to stay all night, if necessary, in order to receive the latest news from Pope, who was at the front, and from McClellan, who was at Alexandria" (118–19). When Pope's telegram was received just before dawn (with the date August 27, 1862, 4:25 a.m.), it speaks of "Intelligence received within twenty minutes" (119) to indicate the timeliness of the information. The telegraph, with its capability of fast message transmission, had a significant impact not only on civilian matters but also on military strategies and commands.

Although many code books mention secrecy, for military purposes it was often critical. The *"BAB" Trench Code* from World War I (FB) has the cover inscription "SECRET: This document is the property of H. B. M. Government and is intended only for the personal information of 'Maj. Hathaway 2nd Bn' [inscribed in handwriting] and of those officers under him whose duties it affects. He is personally responsible for its safe custody, and that its contents are disclosed to those officers and to them only." Further instructions dictate that if the officer has reason to suspect the code has been compromised or fallen into enemy hands, the code should be altered by transpositions of numbers or other manipulations. The necessity for message transmission that could be received within minutes with (relatively) secure encryption is apparent. Under the keyword "Results" is listed the code "139—Our artillery is shelling us" and under "Enemy Forces," "181—Enemy is cutting our wire," a reminder of the conditions under which World War I trench warfare was conducted. Equally indicative of World War I tactics are the entries under "Our Forces": "244—Conditions are favorable for release of gas," "239—all ready for gas attack," and the devastating "256—Gas has blown back." The (NG) copy of *Artillery Code* (1939) has a brown stain on the cover (coffee? blood?), adjacent to the warning "The information given in this document is not to be communicated, either directly or indirectly, to the Press or to any person not holding an official position in His Majesty's Service." Although military forces could not prevent their telegraph lines from being tapped, the code books, as print artifacts, could be protected and kept secure. The problem of communicating over distances while ensuring that only the intended recipients would have access to the messages was solved through the coupling of code and language.

By World War II, "wireless telegraphy," or radio, had become the favored mode of communication. Since radio communications can be intercepted much more easily than tapping telegraph lines, encryption and cryptanalysis were even more crucial. As the successor to telegraphy, radio transmissions played central roles in the Bletchley Park decryption project mentioned earlier and in the less well-known American Enigma decryption project spearheaded by Joseph Desch in Dayton, Ohio. Radio was used more locally as well; the *American Radio Service Code No. 1* (NG) has the cover warning "Not to be taken in front of Brigade Headquarters," presumably to make sure the code book did not advance beyond the secure line of defense. It also has the warning "SECRET: Must Not Fall into Enemy's Hands." Thus the practice of protecting print codes that began with telegraphy continued into the era of radio, with all the possibilities and limitations that implied (for example, the fact that codes were printed meant that they could not easily be changed, although it was always possible to introduce second encryption schemes as discussed below).

The (NCM) *Telegraphic Code Prepared for the Use of Offices and Men of the Navy and Their Families by the Women's Army and Navy League* (n.d. [ca. 1897]) takes into account family situations as well as military exigencies. Such codes as "Accord—You had better start for home as soon as possible" (2); "Adjunct—You had better remain where you are" (2); and "Becloud—Am quite ill. Can you come here at once" (6) illustrate not only the difficulties of separation in wartime but also the expectation that messages could be delivered in minutes or hours. Perhaps because the book was prepared by the Women's Army and Navy League, it shows an inclination toward reporting positive news and avoiding the most devastating outcomes. It lists, for example, "Fable—House was partially destroyed by fire. None of us injured"; "Facial—House was totally destroyed by fire. None of us injured"; and "Factor—House was injured by storm. None of us injured" (14) without giving codes announcing worst-case scenarios of injury or death to family members. The poignant code "Gimlet—Cannot leave until you prepay my passage" (16), as well as the many codes asking that money be telegraphed, bespeak the vulnerabilities of women left without the immediate support of a man in the house, often a necessity for economic survival. They also reveal an underlying confidence that money could be virtualized, sent as a message, and reconstituted as cash at the receiving end.

The *International Code of Signals: American Edition* (1917) shows a dramatic diachronic progression in military-related technologies, reflected in codes initially meant to be conveyed through signal flags (although signal flags predated the telegraph, the construction of signal flag codes was nevertheless influenced by telegraph code books). Initially constructed so that sailors of different nationalities and languages could communicate with one another, this code book is an example of using codes to achieve a sort of universal language (about which more shortly). The 1917 edition, "published by the Hydrographic Office under the authority of the Secretary of the Navy," specifies that the "urgent and important signals are two-flag signals," while those for "general signals" are the more cumbersome (and presumably longer to implement) three-flag signals. (This is analogous to the Morse code system of using the smallest indicator, a single dit, to stand for "e," the most common letter in English messages.) Among the two-flag signals are "FE—Lifeboat is coming to you," as well as the alarming "FG—No boat available."

The military nature of the codes is hinted at in "FM—Allow no communication. Allow no person on board" (37); "FN—Appearances are threatening; be on your guard" (37); "GH—Stranger (or vessel on bearing indicated) is suspicious" (37).

Signal flags were limited to visual contact, and as the century progressed, the letters for signal flags were adapted to Morse code transmission and radio. The *International Code of Signals, as Adopted by the Fourth Assembly of the Inter-Governmental Maritime Consultative Organization in 1965, for Visual, Sound, and Radio Communications, United States Edition* (1968) indicates the simultaneous use of multiple signaling systems. The urgency of some of the plaintext messages indicates the underlying assumption of rapid delivery and fast response. Among the most ominous (conveyed through two-letter codes) are "AJ—I have had a serious nuclear accident and you should approach with caution"; "AK—I have had nuclear accident on board"; and "AD—I am abandoning my vessel which has suffered a nuclear accident and is a possible source of radiation danger" (2-1).

The wide range of topics and situations for which code equivalents were devised indicates how the effects of technogenesis spread. Perhaps only a small percentage were neurologically affected, principally telegraphers and clerks in businesses who routinely sent and received telegrams. Nevertheless, their physico-cognitive adaptations played important roles in making the emergent technologies feasible. Partly as a result of their work, much wider effects were transmitted via the technological unconscious as business practices, military strategies, personal finances, and a host of other everyday concerns were transformed with the expectation of fast communication and the virtualization of commodities and money into information.

Code books had their part to play in these sweeping changes, for they affected assumptions about how language operated in conjunction with code. They were part of a historical shift from inscription practices in which words flowed from hand onto paper, seeming to embody the writer's voice, to a technocratic regime in which encoding, transmission, and decoding procedures intervened between a writer's thoughts and the receiver's understanding of those thoughts. As we have seen, the changes brought about by telegraphy anticipated and partly facilitated the contemporary situation in which all kinds of communications are mediated by intelligent machines. The technogenetic cycles explored in this chapter demonstrate how the connections between bodies and technics accelerated and catalyzed changes in conscious and unconscious assumptions about the place of the human in relation to language and code.

Telegraph code books are also repositories of valuable historical information about attitudes, assumptions, and linguistic practices during the hundred-year span in which they were published. The problem for historians, however, is that they are scattered in different locations and formatted in different ways, so it would take considerable effort to mine them for data that, in many cases, may provide only a few paragraphs of interpretive prose, or maybe a footnote. To remedy this situation, Allen Riddell and I, along with other collaborators, are in the process of constructing a website that can serve as a central portal for telegraph code books, with full-text searching for keywords and associated words and phrases. This has involved digitizing code books in the Gessler collection (see note 1 to the present chapter), as well as creating an interface for other code books already online. In a nice irony, the computer, lineal descendent of the telegraph, provides the functionality that makes access to the historical information contained in telegraph code books easily available: the great-grandson brings the great-grandfather back to life.

Code and the Dream of a Universal Language

Telegraph code books, in addition to offering secrecy, also reflect a desire to create a mode of communication capable of moving with frictionless ease between cultures, languages, and time periods. This tendency appeared first in the guise of polyglot code books that gave code equivalents for phrases listed in multiple languages. As telegraphy made international communication faster and easier than ever before, code books that could function as translation devices became very useful. For example, the (FB) *Tourist Telegraphic Code* by Carlo Alberto Liporace (1929) featured codes with phrase equivalents in several major American and European languages. As seen in figure 5.11, the layout has code words in the rightmost column, with English, German, French, Italian, and Spanish listed in successive columns, with each column showing the appropriate plaintext in the specified language.

The column layout, typical of polyglot code books, is important because it implies that the code mediates between different natural languages. Someone might, for example, encode a telegram referencing the English phrase, reading from the left in the English column across the page to the right side, finding the code word in the rightmost column. An Italian recipient of the telegram would then decode the message locating the code word on the right and reading to the left to find the corresponding Italian phrase. In this sense, code can be seen as a proto–universal language, although neither

CIWHE	We have not gone	Wir sind nicht (nach ...) gefahren	Nous ne sommes pas allés	Non siamo andati	No hemos ido
CIWIE	You » » »	Sie sind » (» ...) »	Vous n'êtes » »	» siete »	» habéis »
CIWOZ	They » » »	(sie) sind » (» ...) »	Ils ne sont » »	» sono »	» han »
CIWRI	**Gondola**	**Gondel**	**Gondole**	**Gondola**	**Góndola**
CIWTO	By gondola	Mit (in) der Gondel	En gondole	Con gondola	Con góndola
CIWUR	Instead of by ... by gondola	Anstatt mit ..., mit der Gondel	Au lieu qu'en ...en gondole	Invece che con ... con gondola	En vez que con ... con góndola
CIWVU	» » » gondola by ...	» » der Gondel, mit ...	» » » gondole en ...	» » » gondola con ...	» » » » góndola, con
CIWXA	One hour of gondola on the Grand Canal	Einstündige Gondelfahrt auf dem Canal Grande	Une heure en gondole sur le Canal Grand	Un'ora con gondola sul Canal Grande	Una hora en góndola por el Canal Grande
CIWYD	Two hours » » » » Grand Canal	Zweistündige » » » » »	Deux heures en gondole sur le Canal Grand	Due ore » » » Canal Grande	Dos horas » » por el Canal Grande
CIXAM	By gondola as far as the Lido	Mit der Gondel bis zum Lido	En gondole jusqu'au Lido	Con gondola fino al Lido	En góndola hasta el Lido
CIXEY	» » » » » » and return	» » » » » » und zurück	Aller et retour jusqu'au Lido en gondole	» » » » » e ritorno	» » » » » y regreso
CIXGE	**Good**	**Gut**	**Bon**	**Buono**	**Bueno**
CIXIK	If ... (it) is good	Wenn (ob) (...) (es) gut ist	Si (...) est bon	Se (...) è buono	Si (...) es bueno
CIXOC	... (it) is good	(Es) ist gut	Est bon	È buono	Es bueno
CIXSC	Very good	Sehr gut	Trés »	Molto buono	Muy bueno
CIXWA	... (it) is not good	(Es) ist nicht gut	N'est pas bon	Non è »	No es »
CIXYG	Not very good	Nicht sehr gut	Pas très »	Non molto buono	» muy bueno
CIYAT	If ... (it) is not good	Wenn (ob) (...) (es) nicht gut ist	Si (...) n'est pas bon	Se (...) non è »	Si (...) no es bueno
CIYCZ	**Government**	**Staat**	**Gouvernement**	**Stato**	**Estado**
CIYEF	Of (from) the (...) Government	Des (...) Staates (vom ... Staat)	Du Gouvernement	Dello (Dallo) Stato	Del Estado
CIYFI	Government employees	Staatsangestellte	Employés du Gouvernement	Impiegati dello »	Empleados del Estado
CIYHO	At Government's expense	Auf Staatskosten (auf Kosten des ... Staates)	Au frais » »	A spese » »	Por cuenta » »

Figure 5.11 Carlo Alberto Liporace's parallel-column layout from *Tourist Telegraphic Code* (1929).

compilers nor users were likely to think of it in this way, partly because the movement back and forth across the page kept code and language on the same level, none diagrammatically more fundamental than any other.

A revolution in language practice, with important theoretical implications, occurred when the conception changed from thinking about encoding/decoding as moving across the printed page to thinking of it as moving up or down between different layers of codes and languages. Impossible to implement in telegraph code books, the idea of layers emerged in the context of computer code, with machine language positioned at the bottom and a variety of programming, scripting, and formatting languages built on it, with the top layer being the screen display that the casual user sees. Positioned at the bottom layer, binary code became the universal symbol code into which all other languages, as well as images and sound, could be translated. Unlike the telegraph code books, which positioned national tongues as the "natural" languages to which "artificial" code groups corresponded, linguistic surfaces in a computer context can be considered as epiphenomena generated by the underlying code. If mathematics is the language of nature, as eighteenth-century natural philosophy claimed, code becomes in this view the universal language underlying all the so-called natural languages.

The ambiguous position of code in relation to "natural" languages (as a generative foundation and as an artificial device used for transmission through technical media), leads to two opposed views of code already implicit in polyglot code books: code makes languages universally legible, and at the same time, code obscures language from unauthorized users. That is, code both clarifies communication and occludes it. John Guillory (2010b), writing about the seventeenth-century interest in code of John Wilkins (another person who dreamed of a rationalized universal language), notes a similar ambiguity: "Wilkins' communication at a great distance is possible only by recourse to the same device–code—that is otherwise the means to *frustrate* communication" (338). In his research into the concept of mediation, Guillory concludes that mediation implies communication over distance. Implicit in his argument is the insight that it is the combination of the two that gives rise to the clarification/occlusion paradox. Successful mediation requires an economical symbol system for legibility (centuries before Morse, Wilkins noted that a binary system can encode any letter or number whatever), while the possibility of interception (because of the distance the message travels) also requires security achieved through cipher schemes—that is, occlusion.

This dual function of code becomes explicit in telegraph code books when compilers instruct users in how to encipher the code groups to achieve greater secrecy than the code book itself allows. For example, two correspondents might agree in advance that in using a five-digit number code, they would create a transposition cipher by starting with the third number in the code group instead of the first (so a code group published in the code book as "58794" would be written instead as "79458"). Telegraphy's overlap with cryptography is one way in which the contraries of a universal language and private secret languages coinhabited the same transmission technology.

The interactions of code and universal language in the twentieth century are located squarely in cryptography. In his famous "Memorandum" (July 1949), Warren Weaver discusses an anecdote drawn from World War II cryptography: an English-speaking cryptanalyst successfully decrypted a message that, unbeknownst to him, had been enciphered from a Turkish original. Without knowing Turkish (or even being able to recognize the decrypted text as Turkish), he nevertheless was able to render the message correctly. Weaver proposes that machine translation might proceed by considering messages in foreign languages as "really" written in English that had been encrypted (the example in his "Memorandum" is Chinese, and in his subsequent essay "Translation" [1955b] Russian—both languages freighted with US anxieties in the cold war era). Machine translation, in this view, becomes a special kind of cryptanalysis. Impressed by the sophisticated techniques used in World War II by the Allies to break the German and Japanese codes, he suggests that these can, in the postwar era, be adapted to machine translation so that international communication, increasingly important as the era of globalization dawned, can be made easy, efficient, and timely. Weaver's essay and "Memorandum" were critiqued at the time by Norbert Wiener, whom Weaver tried to interest in the project, and recently by Rita Raley (2003) and others, who note the imperialist implication that English is the only "real" language to which every other language is subordinate.[10]

Germane for our purposes is another implication of Weaver's proposal: the idea that code can be used not to obscure messages but to render all languages translatable by machines. Hence code goes from being the handmaiden of secrecy to the mistress of transparency. An important (and perhaps inevitable) metaphor in Weaver's argument is the Tower of Babel. In his "Foreword: The New Tower" to Locke and Booth's *Machine Translation of Languages*, Weaver suggests that machine translation should be seen as a "Tower of Anti-Babel" (1955a). In a passage from his "Translation" essay

(1955b), Weaver expands on the idea: "Think, by analogy, of individuals living in a series of tall closed towers, all erected over a common foundation. When they try to communicate with one another, they shout back and forth, each from his own closed tower. It is difficult to make the sound penetrate even the nearest towers, and communication proceeds very poorly indeed. But, when an individual goes down his tower, he finds himself in a great open basement, common to all towers. Here he establishes easy and useful communication with the persons who have also descended from their towers" (23). Note that code, along with the "easy and useful communication" it enables, happens in the basement (at the foundational level), and not, as in Walter Benjamin's essay "The Task of the Translator" (1968a), at the transcendent level, where translatability is guaranteed by the ultimate power of God. For Weaver, code resides in the basement because it functions as the man-made linguistic practice capable of undoing God's disruptive fracturing, at the Tower of Babel, of an ur-language into many languages. In Weaver's analogy, men get back their own not by building the tower higher but by discovering the common basement in which they can communicate freely with one another.

As we have seen, this vision of code as the subsurface lingua franca was realized in a different sense in the development of computer code, especially the layered structures of different software and scripting languages that Rita Raley (2006) has aptly called the "Tower of Languages." Behind this metaphor (or better, this linguistic representation of coding practices) lies the history of early computation. When programming was done by physically plugging jacks into sockets and moving them from one to another to achieve results, as was the case with the earliest digital computers such as ENIAC, metaphors of code layers would be unlikely to occur to anyone, much less comparisons to the Tower of Babel. In "Code.surface || Code.depth" (2006) Raley asks when "do we begin to see people thinking in terms of building layers of abstraction?" (4). Her research into the Association for Computing Machinery proceedings of the 1950s and 1960s indicates that a pivotal point may have been in 1968, when the proceedings of the Garmisch software engineering conference showed "that the notion of multiple layers of software emerges with structured programming and theories of abstract data types" (2006:4). From this point on, different code levels made it possible to translate many different kinds of files—text, sound, image, etc.—into the binary code of machine language. As we know, this and other coding protocols can then be transmitted between machines, where browsers, for example, render pages according to the formatting specified by HTML or XML tags. Although this is not machine translation as such, it represents a version of the lingua franca of which Warren Weaver dreamed.

The World's Most Bizarre Telegraph Code Book

The double vision of code as at once ensuring secrecy and enabling transparency sets the stage for one of the strangest telegraph code books ever written, Andrew Hallner's *The Scientific Dial Primer: Containing Universal Code Elements of Universal Language, New Base for Mathematics, Etc.* (1912). Reflecting a worldview of English as the dominant language, Hallner proposes a universal code into which all languages can theoretically be translated and that also privileges English by inscribing his code equivalents with English phrases and sentences (Hallner also authored *Uncle Sam, the Teacher and Administrator of the World* [1918]). The *Scientific Dial Primer* represents a transition point insofar as its form harkens back to nineteenth-century code books, while its ambitions leap forward to the era of globalized networks of programmable and networked machines.

Hallner conceptualized his "Scientific Dial" as a series of ten concentric rings (see fig. 5.12). The dial is divided into five "Major Divisions" corresponding to the five (English) vowels. Combining different elements and segments of these rings provides an ingenious basis for an entire vocabulary of code words. One-syllable two-letter code words allow 200 permutations, one-syllable three-letter code words allow 2,525 permutations, two-syllable four-letter code words consisting of two vowels and two consonants allow 8,000 permutations, other two-syllable code words with different variations of vowels and consonants allow 12,000 permutations, and five-letter code words create an exponential explosion in which the number of permutations becomes extremely large.

The genius of the Scientific Dial is the rational way in which the various permutations are linked with different classes of plaintext equivalents. For example, the third ring contains numbers from 1 to 20, and the fourth ring contains twenty consonants. Since there are five vowels determining the radial segments (that is, the pie-shaped wedges that constitute the "Major Divisions"), this gives potentially one hundred permutations of the one-syllable consonant-plus-vowel pattern (another hundred can be obtained from the vowel-plus-consonant pattern). The hundred consonant-plus-vowel permutations can be matched with numbers from 1to 100, establishing the

The
SCIENTIFIC DIAL
PRIMER

CONTAINING UNIVERSAL CODE, ELEMENTS OF
UNIVERSAL LANGUAGE, AND NEW BASE
FOR MATHEMATICS

FROM ONE a TO TWENTY

Figure 5.12 Andrew Hallner's Scientific Dial (1912).

code basis for indicating numbers. Similar arrangements are used to denote fractions and measurements of time (based on a decimal arrangement of one hundred seconds to the minute, etc.). These correlations imply that one could, having memorized the Scientific Dial and become proficient in its use, "read" two-letter or even three-letter coded messages directly without having to refer to a code book.

In the author's ambitious (perhaps megalomaniac) projections, the Scientific Dial will be "translated into all the languages of the world, but in them all the *code-words* will remain exactly the same, though the inference or intended meaning, idea, intelligence or information to be conveyed or transmitted is translated" (Hallner 1912:15). He invokes the biblical passage that counters the Tower of Babel, namely the "Pentecost marvel, when all nationalities heard and understood the speaking in their own mother tongue" (15). He imagines that this miracle will be made into as an everyday reality by the Scientific Dial: "You can travel in or through any country, find the way, buy tickets, give orders in hotels and restaurants, attend to toilet, address the barber, arrange your baths, and do anything and everything necessary in travel; and in ordering goods, in exchanging money, and in carrying on general business transactions. And all this knowledge may be acquired in a week! For to acquire and make use of this knowledge is only to understand the Scientific Dial and the principles involved" (16). In this sense, Hallner's vision anticipates the transcodability of machine language in networked and programmable machines, with a crucial difference: unlike strings of ones and zeros, any human anywhere in the world can understand the Scientific Dial codes immediately, whether written or spoken. The place of the human has not yet yielded to the Tower of Languages that would make binary code readable by machines and unreadable by (almost all) people.

Although Hallner does not explicitly refer to cryptography, many circular calculating devices (such as the circular slide rule) and disk cipher devices were widely known in 1912.[11] The connection to cryptography becomes apparent in Hallner's exposition of the different "keys" into which messages may be transposed (see fig. 5.13). Think of the rings in the Scientific Dial not as static entities but as movable devices that can be rotated clockwise or counterclockwise relative to one another. To establish a cipher key, someone sending a message agrees with his recipient beforehand on the direction and magnitude that the rings will be shifted. This provides a way to encode messages that, in the author's view, ensures absolute secrecy, for without knowing the key, no one would be able successfully to decode the message (in actuality even his more complicated schemes for keys would not be difficult

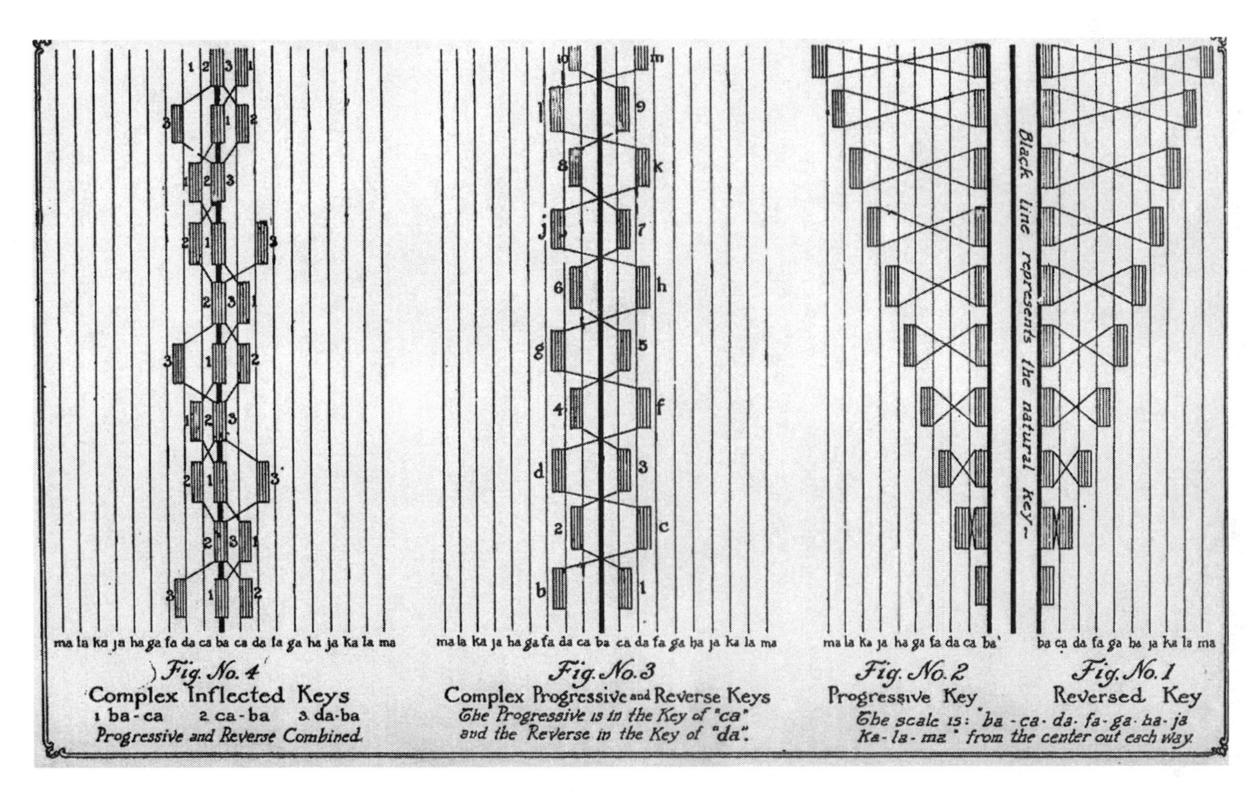

Figure 5.13 Hallner's diagram (1912) for the different "keys" into which code words can be transformed.

to break by modern cryptographic methods, since the transpositions remain regular and predictable throughout the message).

In addition to the universal language created by the Scientific Dial, Hallner understood that additional phrases would be required for social interactions, and these he supplied using four-letter code words (two- and three-letter codes were reserved for the universal language). For the first third of Hallner's text, the code-plaintext equivalents resemble a normal code book (see fig. 5.14 for a sample).

Given the rational form of the Scientific Dial, all the more startling is the text's sudden swerve from single words, phrases, or short sentences for equivalents (such as in fig. 5.14) into long passages that swell to become a grotesque simulacrum of an epistolary novel. At first the passages are anonymous enough conceivably to be useful as code equivalents, although extraordinarily prolix (under "Papa's Kisses" is listed: "ateb = Say, dear, don't forget to kiss . . . . and baby . . . times for me and say to them these are papa's kisses, and have been sent to them by telegraph, and then explain to them what the telegraph is and how it works—won't you, sweetheart, please" (67). The interpenetration of bodies and information discussed earlier can be seen in the notion that "papa's kisses" can be sent as signals along telegraph wires and then rematerialized through the mother's body. Soon, however, the text begins to inscribe plaintext equivalents that refer to specific locations and proper names such as "Floyd" and "Amalia," as well as moralistic messages from a woman to her lover about his bathing habits and his undesirable cigarette habit, along with stern warnings about bars as dens of iniquity.

Particularly startling are the plaintext passages that refer to the book itself. Under "Cer" we read, "Dear Miss Alma: Please accept this copy of the SCIENTIFIC DIAL PRIMER. It contains expressions for feelings and emotions in young people's hearts, and a mode of communication that is direct, pleasing, and strictly private to the parties using it, by resorting to its marvelous arrangement of keys. Let us use it in our confidential correspondence. I suggest the key of 'Ca' as a starter, as that key is easy to remember . . ." (81). A few pages further is "Dcq," presumably the man's response once Alma replies: "Dearest darling, mine: I thank you a thousand times for your sweet, balmy, cordial communication of (Hapelana—June 12th, 1911). I am also much pleased to learn that you appreciate the simple gift of the SCIENTIFIC DIAL PRIMER. 'Simple,' I mean, in the sense of value in dollars and cents. I realize, as you say, that it is a *little marvel*—that Dial—with its manyfold contents and applications . . ." (83). Although many code books include

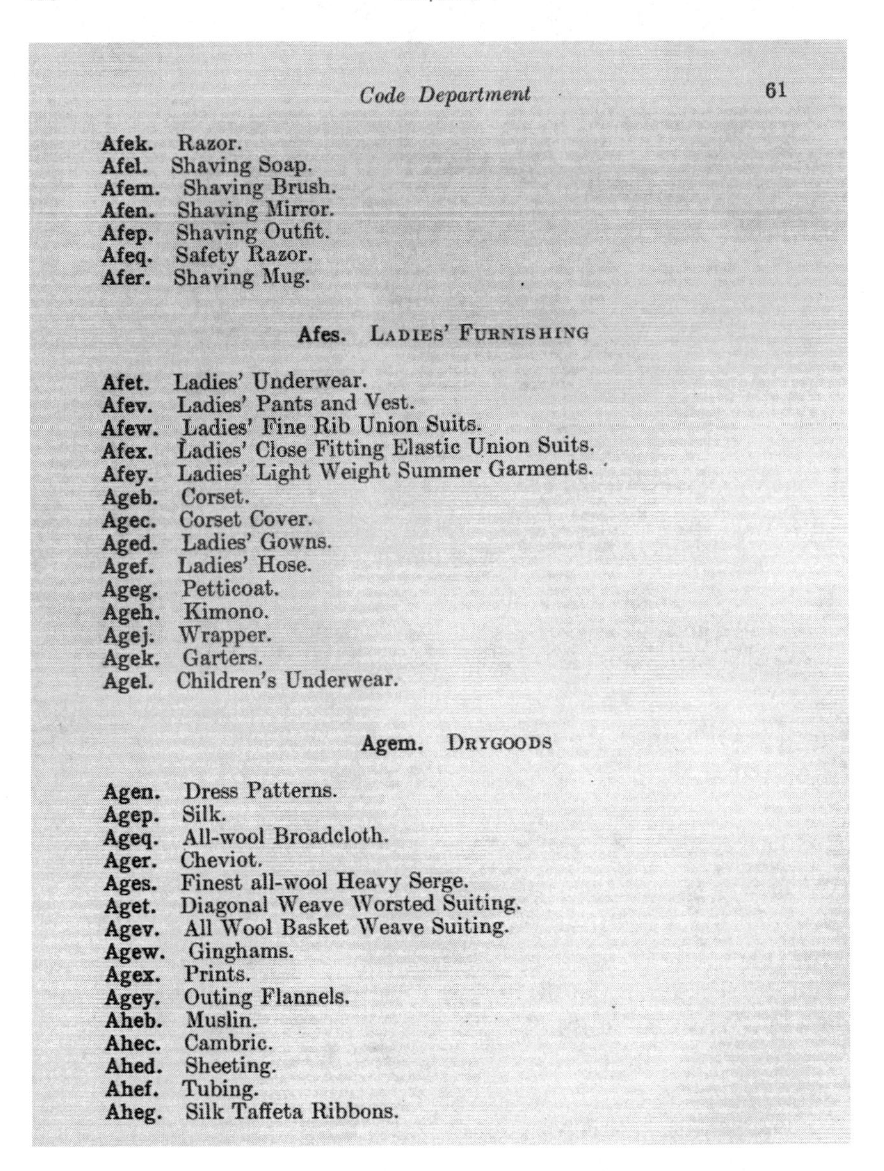

Code Department 61

Afek. Razor.
Afel. Shaving Soap.
Afem. Shaving Brush.
Afen. Shaving Mirror.
Afep. Shaving Outfit.
Afeq. Safety Razor.
Afer. Shaving Mug.

Afes. LADIES' FURNISHING

Afet. Ladies' Underwear.
Afev. Ladies' Pants and Vest.
Afew. Ladies' Fine Rib Union Suits.
Afex. Ladies' Close Fitting Elastic Union Suits.
Afey. Ladies' Light Weight Summer Garments.
Ageb. Corset.
Agec. Corset Cover.
Aged. Ladies' Gowns.
Agef. Ladies' Hose.
Ageg. Petticoat.
Ageh. Kimono.
Agej. Wrapper.
Agek. Garters.
Agel. Children's Underwear.

Agem. DRYGOODS

Agen. Dress Patterns.
Agep. Silk.
Ageq. All-wool Broadcloth.
Ager. Cheviot.
Ages. Finest all-wool Heavy Serge.
Aget. Diagonal Weave Worsted Suiting.
Agev. All Wool Basket Weave Suiting.
Agew. Ginghams.
Agex. Prints.
Agey. Outing Flannels.
Aheb. Muslin.
Ahec. Cambric.
Ahed. Sheeting.
Ahef. Tubing.
Aheg. Silk Taffeta Ribbons.

Figure 5.14 Sample page from first third of *The Scientific Dial Primer* (1912).

praise from users in their prefatory matter, I know of no other book that praises itself in its plaintext phrases!

Interspersed with the epistolary passages are images of bodies that have captions transparently intended to praise the book and to propagate the author's ideology of marital happiness, clean living, and heteronormativity. One image shows a few dozen men and women and several children posed in front of tents in a woodland setting with the caption "A camping community in the mountains using the Scientific Dial code in corresponding with their valley homes" (50). Another is a family portrait of a father (standing) and mother (sitting) along with their four young children, with the caption "A typical happy young family" (79). Yet another shows a group of several young women posed in front of a field of wild grasses with the caption "Ripening into wifehood and motherhood" (84).

What are we to make of the weird combination of a rationalized basis for a universal language, laden with the author's extravagant hopes for worldwide fame and global linguistic transformation, and forays into the particularities of novelistic (and remarkably stilted) passages and images of specific (if anonymous) bodies? It is as if the Scientific Dial, with its purging of the cultural and historical specificities of languages in favor of supposedly neutral "scientific" coding, has provoked a reaction on the author's part that compels him to inscribe, within the code book itself, the cultural and ideological worldview that he evidently regards as the right and true path for human happiness. What the code suppresses, the plaintext sneaks back in through the epistolary passages. What is universal is made excruciatingly particular, and what is rational is subordinated to the emotional raptures of a man and woman in love. Lev Manovich (2002:228) has famously declared that narrative and database are "natural enemies," but here the tabular database of code words and plaintext phrases seems, rather than suppressing narrative, to have catalyzed it into astonishing efflorescence. As for the "universal" future that the author imagines for his book, the only copy I have been able to locate is in the University of California, Berkeley, library—and that copy exists not because the library purchased it but because the author made the library a gift of it.

In the *Scientific Dial Primer*, the tensions between code and language explode into incompatible contraries yoked by violence together, juxtaposed not because they constitute dynamic interplays but simply because the same book covers enclose them. This strangest of telegraph code books shows that the dream of a universal language and "neutral" scientific code is fraught with ideological, political, economic, and social issues that it cannot successfully resolve. If telegraph code books open a window onto "how we [have] thought," this odd specimen reveals that the deep ideological structures and contradictions inhering in linguistic practices cannot be reduced to a decontextualized code passing as a universal language. This is one version of the tension between narrative, with its traditional associations with human agency and theories of mind, and databases as decontextualized atomized information. In linking natural language with codes that became increasingly machine-centric, telegraph code books initiated the struggle to define the place of the human in relation to digital technologies, an agon that remains crucial to the humanities today.

THIRD INTERLUDE

Narrative and Database: Digital Media as Forms

As noted previously, Lev Manovich in *The Language of New Media* has identified narrative and database as competing, and occasionally cooperating, culture forms. His claim that databases are now dominant over narrative seems to me contestable. Nevertheless, I agree with the obvious fact that databases are now pervasive in contemporary society, their growth greatly facilitated by digital media. Databases therefore constitute an important aspect of the technogenetic cycle between humans and technics.

One of the ways in which databases are changing humanistic practices is through GPS and GIS technologies, especially in such fields as spatial history. Whereas qualitative history has traditionally relied on narrative for its analysis, spatial history draws from databases not only to express but also to interrogate historical change. Chapter 6 compares the processual view of space, advocated by geographer Doreen Massey, to the requirements of spatial history projects such as those at the Stanford Spatial History Project and elsewhere.

Whereas Massey desires a view of space as "lively," in sharp contrast to the inert container view of space represented by the Cartesian grid, spatial history researchers want to construct database infrastructures that can be used by other researchers, who can incorporate them into their projects and extend them by adding their own datasets. To ensure interoperability, such infrastructural projects use relational databases keyed to "absolute space," that is, georeferenced to US Geological Survey quads and/or historical maps. Although events that are fundamentally not mappable can be added to relational database tables, the reliance on georeferencing means that many spatial history projects do not (and perhaps cannot) fully achieve the "lively" space that Massey envisions. Since "lively" space presumes the integration of time and space, one possibility is to represent movement (as distinct from static dates) through object-oriented databases. Chapter 6 discusses the different worldviews implicit in relational and object-oriented databases, showing that the latter may incorporate temporal processes in a more integral way than relational databases, notwithstanding that relational databases remain the standard for spatial history projects.

In chapter 7, the narrative-database thread is developed through an analysis of Steven Hall's literary work *The Raw Shark Texts: A Novel* ([2007] 2008a; hereafter *RST*). Presenting first as a print novel, *RST* incorporates other media platforms into its project, including fragments on the Internet, inserted pages in translations of the novel, and other sites, including physical locations. The complete work thus exists as the distributed literary system. One of the text's villains, Mycroft Ward, has transformed into a posthuman subjectivity, called the "Ward-thing" by its antagonists. By separating form from content, a move Alan Liu (2008b) identifies with postindustrial knowledge work, the Ward-thing has become a huge online database, appropriating and acting through "node bodies" after evacuating and absorbing the personality of the original occupant. The Ward-thing thus enables the text to critique database logic and to celebrate narrative aesthetics, especially the creative acts of imagination that turn marks on the page into imagined worlds. In a different sense than Massey intended, fiction can be understood as excelling in making space "lively," a process enacted within the text through the protagonist's journey from ordinary English sites to "unspace" and finally to imaginative space.

This transformation, traditionally associated with immersive fiction, is not an unmixed blessing, as shown by the text's other villain, a "conceptual shark." The shark is the Ludovician, a memory-devouring creature with flesh literally formed from typographical symbols, ideas, and concepts. Both represented by marks and composed of them, the Ludovician embodies the inverse of the Ward-thing's ontology, that is, the complete fusion of form and content. The implications of juxtaposing the Ward-thing and the Ludovician, which can be understood as a battle of database with narrative, surprisingly does not end with the triumph of narrative over database but rather as a more complex negotiation between the powers and dangers of immersive fiction.

Chapter 8 continues the thread of narrative and database through an analysis of Mark Z. Danielewski's *Only Revolutions* (2007b; hereafter *OR*). Whereas *RST* posits narrative in opposition to database, *OR* incorporates database-like information into its literary structure. These two experimental texts thus show the effects of database culture on contemporary literature and speak to the different kinds of responses that contemporary literary texts have to the putative decline of narrative and rise of databases. The effect in *OR* of incorporating database elements into the page surface is to tilt the text toward a spatial aesthetic. Similar to relational database tables displayed on a two-dimensional plane, the pages of *OR* inscribe complex spatial symmetries that express various combinations of a 360-degree "revolution." The text's topographic surfaces are also in dynamic interplay with its volumetric properties as a three-dimensional object. The material effects of spatiality are densely interrelated with the text's bibliographic codes, especially typography, narrative placement, and sidebar chronologies, as well as with its semantic and narrative content. The result creates a "readable object" with a high dimensionality of potential reading paths that far exceeds the two and three dimensions of its spatiality in a conventional sense.

In addition to its spatial aesthetic, *OR* is written under a severe set of constraints that position the text as the mirror inversion of Danielewski's previous best-selling novel *House of Leaves* (2000). Because of the constraints, *OR* demonstrates extensive patterning in narrative parallels and other devices. It thus lends itself to machine analysis, which excels in detecting patterns. Machine reading, a topic introduced in chapters 2 and 3, is here applied to the tutor text of *OR*. The analysis, coauthored with Allen Beye Riddell and explained in a "coda" to chapter 8, reveals how extensive is the patterning. It also creates a way to discover exactly what words (as distinct from general concepts) are forbidden to appear in the text. The coda thus makes good the appeal for an expanded repertoire of reading practices articulated in chapter 3, as well as to a Comparative Media Studies approach that undergirds the comparison of narrative and database in contemporary fiction.

6

Narrative and Database

Spatial History and the Limits of Symbiosis

Ah, the power of metaphors—especially when a metaphor propagates with viral intensity through a discursive realm. At issue is Lev Manovich's characterization of narrative and database in *The Language of New Media* (2002) as "natural enemies" (228), a phrase Ed Folsom rehearses in his discussion of the *Walt Whitman Archive* (2007). The metaphor resonates throughout Folsom's essay in phrases such as "the attack of database on narrative" (1574), culminating in his figure of database's spread as a viral pandemic that "threatens to displace narrative, to infect and deconstruct narrative endlessly, to make it retreat behind the database or dissolve back into it" (1577). In this imagined combat between narrative and database, database plays the role of the Ebola virus, whose voracious spread narrative is helpless to resist. The inevitable triumph of database over narrative had already been forecast in Manovich's observation that "databases occupy a significant, if not the largest, territory of the new media landscape" (2002:228). Indeed, so powerful and pervasive are databases in Manovich's view that he finds it "surprising" narratives continue to exist at all in new media (228). In Manovich's view, the most likely explanation of narrative's persistence is the tendency in new media to want to tell a story, a regressive tendency he identifies with cinema. Even this, he suggests, is in the process of being eradicated by experimental filmmakers such as Peter Greenaway (237–39).

Rather than being natural enemies, narrative and database are more appropriately seen as *natural symbionts*. Symbionts are organisms of different species that have a mutually beneficial relation. For example, a bird picks off bugs that torment a water buffalo, making the beast's existence more comfortable; the water buffalo provides the bird with tasty meals. Because database can construct relational juxtapositions but is helpless to interpret or explain them, it needs narrative to make its results meaningful. Narrative, for its part, needs database in the computationally intensive culture of the new millennium to enhance its cultural authority and test the generality of its insights. If narrative often dissolves into database, as Folsom suggests, database catalyzes and indeed demands narrative's reappearance as soon as meaning and interpretation are required (as the discussion at the conclusion of chapter 5 illustrates). The dance (or as I prefer to call it, the complex ecology) of narrative and database has its origins in the different ontologies, purposes, and histories of these two cultural forms. To understand more precisely the interactions between these two cultural forms, let us turn now to consider their characteristics.

The Different Worldviews of Narrative and Database

As Manovich (2002) observes, database parses the world from the viewpoint of large-scale data collection and management. For the late twentieth and early twenty-first centuries, this means seeing the world in terms that the computer can understand. By far the most pervasive form of database is the relational, which has almost entirely replaced the older hierarchical, tree, and network models and continues to hold sway against the newer object-oriented models. In a relational database, data are parsed into tables consisting of rows and columns, where the column heading, or attribute, indicates some aspect of the table's topic. Ideally, each table contains data pertaining to only one "theme" or central data concept. One table, for example, might contain data about authors, where the attributes might be last name, first name, birth date, death date, book titles, etc.; another might have publishers' data, also parsed according to attributes; another, books. Relations are constructed among data elements in the tables according to set-theoretic operations, such as "insert," "delete," "select," and especially "join," the command that allows data from different tables to be combined. Common elements allow correlations to be made between tables; for example, Whitman would appear in the authors table as an author and in the books table correlated with the titles he published; the publishers table would correlate with the books table through common elements, and through these elements back to the authors table. Working through these kinds of correlations, set-theoretic operations allow new tables to be constructed from existing ones. Different interfaces can be designed according to the particular needs of users. Behind the interface, whatever its form, is a database-management system that employs set-theoretic notation to query the database and manipulate the response through SQL (SQL is commonly expanded as structured query language).

The great strength of database, of course, is the ability to order vast arrays of data and make them available for different kinds of queries. Two fundamental aspects typically characterize relational databases. One, indicated above, is their construction of relations between different attributes and tables. The other is a well-constructed database's self-containment, or, as the technical literature calls it, self-description. A database is said to be self-describing because its user does not need to go outside the database to see what it contains. As David Kroenke and David Auer put it in *Database Concepts* (2007), the "structure of the database is contained within the database itself" (13), so that the database's contents can be determined just by looking inside it. Its self-describing nature is apparent in SQL commands. For the database mentioned above containing information about authors, books, and publishers, for example, a typical SQL command might take the generalized form "SELECT AUTHOR.AuthorName, BOOK. BookTitle, BOOK.BookDate, BOOK.Publisher, PUBLISHER.Location," where the table names are given in all capitals (as are SQL commands) and the data records are categorized according to the attributes, with a period separating table name from attribute. The database's self-description is crucial to being able to query it with set-theoretic operations, which require a formally closed logical system upon which to operate. This is also why databases fit so well with computers. Like databases, computers work according to the operations of formal logic as defined by the logic gates that underlie all executable commands.

The self-describing nature of database provides a strong contrast with narrative, which always contains more than indicated by a table of contents or a list of chapter contents. Databases can, of course, also extend outward when they are linked and queried as a network—for example, in data mining and text mining techniques—but they do not lose the formal properties of closure that make them self-describing artifacts. Nevertheless, the technologies of linking databases have proven to be remarkably powerful, and the relations revealed by set-theoretic operations on networks of linked databases can have stunning implications. For example, data- and text-mining techniques allowed epidemiologist researchers Don Swanson and N. R. Smalheiser (1994, 1997) to hypothesize causes for rare diseases that hitherto had resisted analysis because they occurred infrequently at widely separated locales. Even in this case, however, the meaning of the relations posited by the database remains outside the realm of data techniques. What it means that Whitman, say, used a certain word 298 times in *Leaves of Grass* while using another word only 3 times requires interpretation—and interpretation, almost inevitably, invokes narrative to achieve dramatic impact and significance. Many data analysts and statisticians are keenly aware of this symbiosis between narrative and data. John W. Tukey, in his classic textbook *Exploratory Data Analysis* (1997), for example, explains that the data analyst "has to learn . . . how to expose himself to what his data are willing—or even anxious—to tell him" (21), following up the lesson by later asking the student/reader "what story did each [dataset] tell" (101).

Database and narrative, their interdependence notwithstanding, remain different species, like bird and water buffalo. Databases must parse information according to the logical categories that order and list the different data elements. Indeterminate data—data that are not known or otherwise elude the boundaries of the preestablished categories—must either be represented through a null value or not be represented at all. Even though some relational databases allow for the entry of null values, such values work in set-theoretic operations as a contaminant, since any operation containing a null value will give a null value as its result, as multiplying a number by zero yields zero. Null values can thus quickly spread through a database, rendering everything they touch indeterminate. Moreover, database operations say nothing about how data are to be collected or which data should qualify for collection, nor does it indicate how the data should be parsed and categorized. Constructing a database always involves assumptions about how to set up the relevant categories, which in turn may have ideological implications, as Geoffrey C. Bowker and Susan Leigh Star have observed (2000). In addition, such decisions greatly influence the viability, usefulness, and operational integrity of databases. Thomas Connolly and Carolyn Begg in

Database Systems (2002) estimate that for corporate database software development projects, 80 to 90 percent do not meet their performance goals, 80 percent are delivered late and over budget, and 40 percent fail or are abandoned (270). Anticipating such problems, database textbooks routinely advise students to keep the actual design of the database a closely guarded secret, confining discussions with the paying client to what the interface should look like and how it should work. In addition to not "confusing" the client, such a strategy obscures suboptimal performance.

The indeterminacy that databases find difficult to tolerate marks another way in which narrative differs from database. Narratives gesture toward the inexplicable, the unspeakable, the ineffable, whereas databases rely on enumeration, requiring explicit articulation of attributes and data values.[1] While the concatenation of relations might be suggestive, as the epidemiology example illustrates, databases in themselves can only speak that which can explicitly be spoken. Narratives, by contrast, invite in the unknown, taking us to the brink signified by Henry James's figure in the carpet, Kurtz's "The horror, the horror," Gatsby's green light at pier's end, Kerouac's beatitude, Pynchon's crying of Lot 49. Alan Liu, discussing the possibilities for this kind of gesture in a postindustrial information intensive era, connects it with "the ethos of the unknown" and finds it expressed in selected artworks as a "data pour," an overflowing uncontainable excess that he links with transcendence (2008b:81).

Whereas database reflects the computer's ontology and operates with optimum efficiency in set-theoretic operations based on formal logic, narrative is an ancient linguistic technology almost as old as the human species. As such, narrative modes are deeply influenced by the evolutionary needs of humans negotiating unpredictable three-dimensional environments populated by diverse autonomous agents. As Mark Turner has argued in *The Literary Mind: The Origins of Thought and Language* (1998), stories are central in the development of human cognition. Whereas database allows large amounts of information to be sorted, cataloged, and queried, narrative models how minds think and how the world works, projects in which temporality and inference play rich and complex roles. Extending Paul Ricoeur's (1990) work on temporality and Gérard Genette's (1983) on narrative modalities, Mieke Bal (1998) analyzes narrative as requiring, at a minimum, an actor and narrator and consisting of three distinct levels, text, story, and fabula, each with its own chronology (6). To this we can add Brian Richardson's emphasis in *Unlikely Stories: Causality and the Nature of Modern Narrative* (1997) on causality and inference in narrative.[2]

Why should narrative emphasize these aspects rather than others? Bound to the linear sequentiality of language, narrative complicates it through temporal enfoldings of story (or, as Genette prefers to call it, discourse) and fabula, reflecting the complexities of acting when knowledge is incomplete and the true situation may be revealed in an order different from the one logical reconstruction requires. Narrator and actor inscribe the situation of a subject constantly negotiating with agents who have their own agendas and desires, while causality and inference represent the reasoning required to suture different temporal trajectories, motives, and actions into an explanatory frame. These structures imply that the primary purpose of narrative is to search for meaning, making narrative an essential technology for humans, who can arguably be defined as meaning-seeking animals.

Bound to the linear order of language through syntax, narrative is a temporal technology, as the complex syncopations between story and fabula demonstrate. The order in which events are narrated is crucial, and temporal considerations are centrally important in narratology, as Ricoeur's work (1990), among others, illustrates. Data sets and databases, by contrast, lend themselves readily to spatial displays, from the two-dimensional tables typical of relational databases to the more complex *n*-dimensional arrays and spatial forms that statisticians and data analysts use to understand the stories that data tell.

Manovich touches on this contrast when he argues that for narrative, the syntagmatic order of linear unfolding is actually present on the page, while the paradigmatic possibilities of alternative word choices are only virtually present. For databases, the reverse is true: the paradigmatic possibilities are actually present in the columns and rows, while the syntagmatic progress of choices concatenated into linear sequences by SQL commands is only virtually present (Manovich 2002:228) This influential formulation, despite its popularity, is seriously flawed, as Allen Bye Riddell points out (pers. communication April 7, 2008). Recall that in semiotics, the alternative choices of the paradigm interact with the inscribed word precisely because they are absent from the page, although active in the reader's imagination as a set of related terms. Contrary to Manovich's claim, databases are *not* paradigmatic in this sense. In a relational database configured as columns and rows, the data values of a row constitute the attributes of a given record, while the columns represent the kinds of attribute itemized for many different records. In neither the rows nor columns does a logic of substitution obtain; the terms are not synonyms or sets of alternative terms but different data values.

These objections notwithstanding, Manovich's formulation contains a kernel of truth. Search queries, as we have seen, allow different kinds of attributes and data values to be concatenated. The concatenated values, although not syntagmatic in the usual sense, can be seen as a (virtual) temporal process in which it is not syntax but the SQL command that determines the concatenation's order. Thus, although Manovich's formulation is not technically correct, it captures the overall sense that the temporal ordering crucial for narrative is only virtually present in the database, whereas spatial display is explicit. I would add to this the observation that time and space, the qualities Kant identified as intrinsic to human sensory-cognitive faculties, inevitably coexist. While one may momentarily be dominant in a given situation, the other is always implicit, the natural symbiont whose existence is inextricably entwined with its partner. It should be no surprise, then, that narrative and database align themselves with these partners, or that they too exist in symbiosis with one another.

Given this entwinement, is it plausible to imagine, as Manovich and Folsom imply at various points, that database will replace narrative to the extent that narrative fades from the scene? A wealth of evidence points in the other direction: narrative is essential to the human lifeworld. Jerome Bruner, in his book significantly entitled *Acts of Meaning: Four Lectures on Mind and Culture* (1992), cites studies indicating that mothers tell their children some form of narrative several times each hour to guide their actions and explain how the world works (81–84). We take narrative in with mother's milk and practice it many times every day of our lives—and not only in high culture forms such as print novels. Newspapers, gossip, math story problems, television dramas, radio talk shows, and a host of other communications are permeated by narrative. Wherever one looks, narratives surface, as ubiquitous in everyday culture as dust mites.

What has changed in the informative-intensive milieu of the twenty-first century is the position narrative occupies in the culture. Whereas in the classical Greek and Roman era narrative was accepted as an adequate explanation for large-scale events—the creation of the world, the dynamics of wind and fire, earth and water—global explanations are now typically rooted in data analysis. If we want to understand the effects of global warming or whether the economy is headed for a recovery, we likely would not be content with anecdotes about buttercups appearing earlier than usual in the backyard or Aunt Agnes's son finally finding a job. Data, along with the databases that collect, parse, and store them and the database-management systems that concatenate and query them are essential for understanding large-scale phenomena. At the global level, databases are essential. Even there, however, narrative enters in the interpretation of the relations revealed by database queries. When Ben Bernanke testifies before Congress, he typically does not recount data alone. Rather, he tells a *story*, and it is the story, rather than the data by themselves, that propagates through the news media, because it encapsulates in easily comprehensible form the meaning exposed by data collection and analysis.

In contrast to global dynamics, narrative at the local level remains pervasive, albeit more and more infused by data. In the face of the overwhelming quantities of data that database-management systems now put at our fingertips, no one narrative is likely to establish dominance as *the* explanation, for the interpretive possibilities proliferate as databases increase. In this respect, the advent of the Internet, especially the World Wide Web, has been decisive. Never before in the history of the human species has so much information been so easily available to so many. The constant expansion of new data accounts for an important advantage that relational databases have over narratives, for new data elements can be added to existing databases without disrupting their order. Unlike older computer database models in which memory pointers were attached directly to data elements, relational databases allow the order of the rows and columns to vary without affecting the system's ability to locate the proper elements in memory. This flexibility allows databases to expand without limitation (subject, of course, to the amount of memory storage allocated to the database). Narrative in this respect operates quite differently. Sensitively dependent on the order in which information is revealed, narrative cannot in general accommodate the addition of new elements without, in effect, telling a different story. Databases tend toward inclusivity, narratives toward selectivity. Harry Mathews explores this property of narrative in *The Journalist: A Novel* (1994), where the unnamed protagonist, intent on making a list of everything that happens in his life, thinks of more and more items, with the predictable result that the list quickly tends toward chaos as the interpolations proliferate. The story of this character's life cannot stabilize, because the information that constitutes it continues to grow exponentially, until both list and subject collapse.

That novels like *The Journalist* should be written in the late twentieth century speaks to the challenges that database poses to narrative in the age of information. No doubt phenomena like this explain why Manovich would characterize database and narrative as "natural enemies" and why thoughtful scholars would propagate the metaphor. Nevertheless, the same dynamic also explains why the expansion of database is a powerful force constantly spawning new narratives. The flip side of narrative's inability to tell *the* story is the proliferation of narratives as they transform to accommodate new data and mutate to probe what lies beyond the exponentially expanding infosphere. No longer singular, narratives remain the necessary others to database's ontology, the perspectives that invest the formal logic of database operations with human meanings and gesture toward the unknown hovering beyond the brink of what can be classified and enumerated.

Spatial History: A Field in Transition

Among the disciplines that routinely rely on narrative is qualitative history. Traditionally seen as centering on "change through time," history has recently been going through a "spatial turn," in which databases, GIS, and GPS technologies have provided an array of tools for rethinking and re-representing the problematics of history in spatial terms. Accordingly, historians have moved into alliance with geographers in new ways. This movement has its own tensions, however, for geographers often rely on discursive methods of explanation, whereas historians are turning more and more to databases.

The tension can be seen in the work of Doreen Massey (2005) compared to such database history projects as those of the Stanford Spatial History Project. Massey, an important contemporary geographer, has a dream: to replace the idea of space as an inert container with a conceptualization of it as an emergent property constructed through interrelations and containing diverse simultaneous trajectories. In general, she advocates a view of space as "lively" rather than "dead." Above all, she does not want space conceptualized as a pre-given Cartesian manifold that extends in every direction, infinitely subdividable and homogeneous through multiple scale values. Yet this seems to be precisely what Google Earth, GIS, and GPS digital technologies offer, with the illusion of present tense rendering, zoom functions through multiple scale levels, and seamless transitions between map, satellite, and street views. Given the power, pervasiveness, and allure of such visualizations, can Massey's dream gain traction in contemporary representations? More generally, does the "spatial turn" in digital history projects imply that the traditional focus on time for historians has now been transformed into spatializations that take the Cartesian grid as the basis for historical representations? Answering these questions requires a deeper understanding of what Massey's dream implies, along with its roots in prior work by geographers and an analysis of how GIS and GPS technologies are actually being used in specific spatial history projects. As we will see, the answers are less clear-cut than a simple opposition between Cartesian and "lively" space may suggest.

Roots of a Dream

Understanding Massey's dream requires rewinding to Henri Lefebvre's *The Production of Space* ([1974] 1992). Arguing against reducing space to "the level of discourse" (7), Lefebvre maintained that "real" space is constructed through social practice and that the "logico-mathematical" space of the Cartesian grid constantly interacts with social space: "Each of these two kinds of space involves, underpins and presupposes the other" (14). Working through the implications of this conception, Lefebvre proposed a triad of terms: spatial practices, which produce space as perceived within a society; representations of space, tied to spatial signs and codes and resulting in space as conceived, as in architectural drawings, for example; and representational space, which is space "as directly lived through its associated images and symbols" (38–39), such as a cathedral or railway terminal.

Lefebvre's work sparked a revolution in geography, with theorists and practitioners delineating precisely how social practices construct social spaces, for example in the regional studies done by Nigel Thrift and Ash Amin, which posited regions not merely as responding to changes wrought nationally and internationally but generating change in themselves (Amin and Thrift, 2002; Thrift 1996). Also in this vein are Massey's studies of regional geographies, such as "Uneven Development: Social Change and Spatial Divisions of Labour" (1994b), which show that regions develop distinctive spatial practices with implications for other kinds of social structures such as class and gender. Complicating the emphasis on regions was Yi-Fu Tuan's influential *Space and Place: The Perspective of Experience* (1977), a seminal text in the development of humanistic geography that focuses on human consciousness and its ability to attribute meaning to places. Distinguishing "place" from "space," Tuan suggests that "place," redolent with lived memories and experience, offers stability and security, while "space," more abstract than place, offers "openness, freedom, and threat" (6). Although Tuan himself noted that " 'space' and 'place' require each other for definition," subsequent work in humanistic geography tended to privilege place over space, primarily because its associations with lived experience tied it more closely to consciousness than did the abstractions of space. For their part, many Marxist geographers such as Edward Soja and David Harvey strenuously objected to the emphasis on consciousness in humanistic geography, preferring to focus on the primacy of capitalist relations in structuring economic and social life. In response to Manuel Castells's argument (1996) that urban areas, with increased communication possibilities brought about by digital technologies, had become "spaces of flows," some geographers vigorously debated this formulation, emphasizing the embeddedness and embodied nature of interactions with the environment, even while also emphasizing the importance of electronic communication among objects (Thrift 2005).

All this work contributed to Massey's argument in "A Global Sense of Place"(1994a) that geographers need to develop a progressive sense of place that avoids the reactionary trap of seeing "place" as "a response to desire for fixity and for security of identity in the middle of all the movement and change. . . . On this reading, place and locality are foci for a form of romanticized escapism from the real business of the world" (151). To reposition place while still acknowledging the social need for a sense of rootedness, Massey proposes a view of place that sees its specificity as resulting "not from some long internalized history but the fact that it is constructed out of a particular constellation of social relations, meeting and weaving together at a particular locus" (154). Satirizing viewpoints that make "ritualistic connections to 'the wider system'—the people in the local meeting who bring up international capitalism every time you try to have a discussion about rubbish-collection," she nevertheless insists "there are real relations with real content—economic, political, cultural—between any local place and the wider world in which it is set" (155). The key here is conceptualizing place not as a fixed site with stable boundaries but rather as a dynamic set of interrelations in constant interaction with the wider world, which nevertheless take their flavor and energy from the locality they help to define. In addition to emphasizing the dynamic nature of social space, interrelations are important for Massey because they help to overcome the local/global binary. Since these relations are always in flux (both within and between the local and the global), it follows as a necessary consequence that space is emergent, always open and never predictable in its future configurations. In this sense, it forms a strong contrast with the static and infinite predictability of Cartesian space.

Massey's insistence on the importance of diverse simultaneous trajectories has a political as well as a theoretical purpose. She wants to disrupt narratives that place first world and third world nations on the same universal trajectory, a construction implying that "developing" nations will eventually reach the same point as "developed" countries; they are not different in kind but only further back along the arc of progress. Criticizing conceptions of space typical of modernity, she writes, "Different 'places' were interpreted as different stages in a single temporal development. [In] all the stories of unilinear progress, modernization, development, the sequence of modern modes of production . . . Western Europe is 'advanced,' other parts of the world 'some way behind,' yet others are 'backward.' Africa is not *different* from Western Europe, it is (just) behind" (2005:68; emphasis in original). Not only is space co-opted into a homogeneous grid, but temporality is compromised as well: "In these conceptions of singular progress (of whatever hue), temporality itself is not really open. The future is already foretold; inscribed into the story" (68). The alternative is what Massey, following Johannes Fabian, calls "coevalness," which "concerns a stance of recognition and respect in situations of mutual implication. It is an imaginative space of engagement" (70). Making space "lively," then, implies not only that it is emergent but also that multiple temporal trajectories inhabit it in diverse localities across the globe.

Tensions between the Dream and Spatial History Projects

However attractive these propositions sound in the abstract, the question is how to accomplish them in spatial histories and other kinds of spatial projects. Here Massey goes from the exemplary to the opaque. She rightly notes that there has been a long association "between space and the fixation of meaning," noting that representation "has been conceptualized as spatialization" (2005:20). Referencing Bergson, she notes that "*in the association of [space] with representation* it was deprived of dynamism, and radically counterpoised to time" (21; emphasis in original). Arguing that history cannot be derived "from a succession of slices through time" (22), she detects in Bergson's arguments a slide "from spatialisation as an activity to space as a dimension" (23). As a result, space is "characterized as the dimension of quantitative divisibility. . . . This is fundamental to the notion that representation is spatialisation" (23). She argues, on the contrary, for the interpenetration of space and time and for the conceptualization of space "as the dimension of multiple trajectories, simultaneity of stories-so-far. Space as the dimension of a multiplicity of durations" (24). Representation in this view becomes "not the spatialisation of time (understood as the rendering of time as space) but the representation of time-space. What we conceptualise . . . is not just time but space-time" (27). Time in this view rescues space

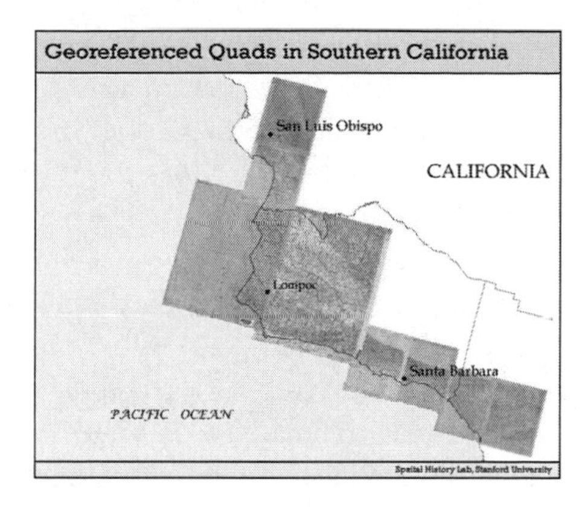

Figure 6.1 Georeferenced quads for Southern California. From Ryan Delaney's "Backend Visualizations: Tools for Helping the Research Process." The Spatial History Project (August 2009), http://www.stanford.edu/group/spacialhistory/cgi-bin/site/pub.php, accessed August 30, 2009. Copyright © Stanford University. All rights reserved.

from being static, while "lively" space rescues time from being flattened into nonexistence by the Cartesian grid.

We can measure these aspirations against the choices that actual projects in spatial history have made. The Spatial History Project at Stanford University, directed by Richard White, offers several well-crafted examples (http://www.stanford.edu/group/spatialhistory/cgi-bin/site/index.php). The Western Railroads project (Shnayder 2009) began constructing its site by matching US Geological Survey (USGS) quads to 3-D Google images, using polynomial functions to correct for corners that did not quite match[3] and importing them into ArcGIS[4] (fig. 6.1). This provided the team with survey information that included reliable metadata correlated with on the ground mapping. With this basic infrastructure in place, they then created relational databases containing the station names, railroad routes, lengths of rail between stations, etc. These tables they thought of as the "hardware" for the project, the historical data that could be matched directly with GIS maps and positions. Their next step was to build more databases containing nonspatial information such as construction dates, distances between stations, freight rates, accidents, etc.; this they thought of as the "software" that ran on the "hardware" of georeferenced spatial data. For example, the construction dates can be correlated with the railroad routes to create visualizations showing the growth of a particular railroad line over time, a procedure integrating time and space. The third step was to build relational connections in database form that allowed the spatial data to be correlated with the nonspatial data, for example, "ConnectionsHaveDistances," bridging the data on connections to distance information, and "ConnectionHas-FreightRates," bridging the data on connections to the freight rates for a specific connection.

From the viewpoint of Massey's dream, these procedures are flawed because at the bottom of the correlations is the georeferencing that maps the USGS quads onto what Richard White (2010) calls "absolute space," space measured in miles, kilometers, etc. (para. 25). The ability to correlate social practices (in this case, building railroads) to absolute space has a powerful advantage that Massey does not acknowledge, namely the creation of an infrastructure that allows databases from the project to be used in novel ways by researchers who have other data sets that they can combine with the existing databases of the Spatial History Project. The researchers at Stanford emphasize that they intend to lay "the foundation for any number of future research projects" (Shnayder 2009:para. 3), pointing out that "the Western Railroads Geodatabase has also allowed us to increase researcher accessibility to our data within the Lab and will eventually allow outside researchers and interested individuals to gain access to our work as well. The geodatabase allows several researchers to simultaneously edit or draw features, which in turn speeds up the digitization or analysis process" (Shnayder 2009). Having absolute space as a fundamental reference point is essential to allowing connections to be made between new research questions and the infrastructure created by this project.

What comes into view with this observation are the crucial differences between discursive research of the kind Massey and other geographers practice, and the database projects of the Stanford Spatial History Project and similar database-driven projects. While a discursive practitioner may hope that her ideas will be picked up by others, and perhaps summaries of her data as well, these ideas circulate in the form of what Richard Dawkins called memes, ideas that follow evolutionary dynamics but replicate in human brains rather than through sexual or asexual reproduction. Memetic reproduction can be a powerful disseminative force (think, for example, of Agamben's "bare life," Foucault's "biopolitics," Deleuze and Guattari's "rhizome"), but they are primarily semiotic rather than data driven, circulating through rhetorical formulations rather than database structures that can be linked directly with other researchers' data sets. Memes do not allow new insights to emerge as a result of establishing correlations between existing databases and new datasets; rather, they serve as catalysts for reframing and reconceptualizing questions that are given specific contexts by the writer appropriating the memes for his or her own purposes. In this sense, memes contrast with what Susan Blackmore (2010) calls temes, ideas that replicate through technical objects rather than human brains. A database is of course originally constructed through human cognition, but once built, it carries its implications forward by being connected to other data sets. It becomes, that is, a teme, a technical object through which assumptions replicate. When databases function as the infrastructure for other projects, as do the georeferenced USGS quads in the Western Railroads project, they can carry implicit assumptions into any number of other projects, in particular the assumption that spatial practices are built on top of absolute space.

Complicating this assumption is the GIS functionality of creating layers that indicate relations between different kinds of parameters. An example is Zephyr Frank's "Terrain of History" project at the Stanford Spatial History Project, which creates layers on top of a historical base map of Rio de Janeiro in the 1840s–1870s showing the topography of the city's hills. The next layer after the georeferenced historical map is a digital street map with geocoded addresses, followed by a map showing property value contours, built by geocoding the addresses of urban property values and creating an interpolated surface. Extruding these values into three dimensions allows correlation between property values and the topographic map of the city. Here interrelationality emerges through the correlations between different kinds of layers, as well as between points within the same layer, such as the spatial distribution of property values relative to one another.

A further complication emerges from the practice of creating "distorted" maps that show how absolute space is transformed/deformed in relation to cultural or economic factors, thus instantiating Lefebrve's point about social practices creating social spaces. An example drawn from my own experience might show, for example, the relative distance from the San Fernando valley to downtown Los Angeles as a function of time of day. Although the absolute distance remains the same, the valley is much farther from downtown at 9:00 a.m. than it is at 9:00 p.m. Richard White and his collaborators on the Western Railroads project have developed a Flash implementation that shows this effect with freight charges as a function of distances in 1876 (figs. 6.2 and 6.3). These images were obtained by correlating the database field of freight charges with track length. As White observes, it is difficult to understand the relationship between the two solely by studying a table of freight charges. In the Flash visualization, however, the disproportionality quickly becomes clear and, White suggests, indicates why wheat and

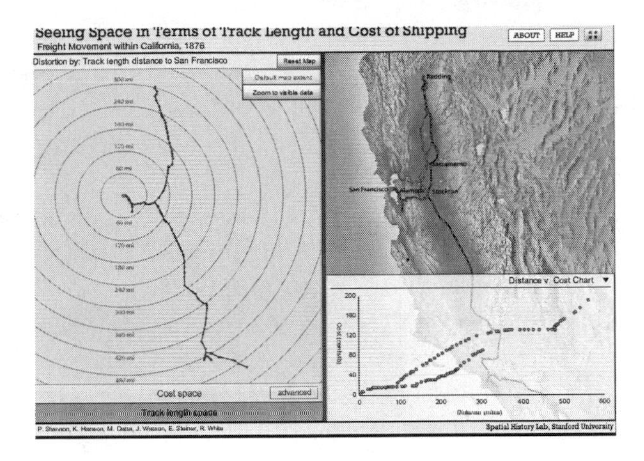

Figure 6.2. Cost-versus-distance representations for freight charges. Track length space is shown on left. From Ryan Delaney's "Backend Visualizations: Tools for Helping the Research Process." The Spatial History Project (August 2009), http://www.stanford.edu/group/spacialhistory/cgi-bin/site/pub.php, accessed August 30, 2009. Copyright © Stanford University. All rights reserved.

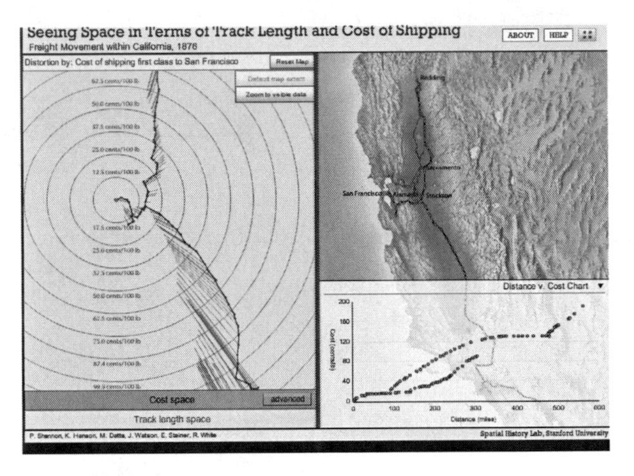

Figure 6.3 Track length is distorted when mapped as a function of cost (gray shaded lines trace the "undistorted" correlation shown in fig. 6.2). From Ryan Delaney's "Backend Visualizations: Tools for Helping the Research Process." The Spatial History Project (August 2009), http://www.stanford.edu/group/spacialhistory/cgi-bin/site/pub.php, accessed August 30, 2009. Copyright © Stanford University. All rights reserved.

grain farmers were so angry with the railroads during this period. Of course, verbally calling this image a "distortion" and visually referencing it to a georeferenced image of track length gestures toward the notion of space as absolute, even while also bringing into view how economic practices construct the socially mediated space of freight charges.

The Cultural and Historical Specificity of Interrelationality

Massey argues that making interrelationalty a key feature of space will ensure that the modernist trajectory of progress she criticizes will not be allowed to dominate. The ways in which interrelationality can define space by using a georeferenced ArcGIS system is illustrated by Peter Bol's "Creating a GIS for the History of China" (2008). He observes that in the more than two thousand years covered by the China Historical GIS project, the dominant conception of space for hundreds of years was not of parcels marked by clear-cut boundaries but rather hierarchical networks of points, with the court marking the central node from which authority, wealth, and administration radiated outward through a series of nodes cascading through smaller and smaller administrative levels, all the way down to villages and individual landholders. Although relationality characterizes the practices that constructed this conception of space, the relations were not of the same kind or significance: orders and administrative regulations flowed from the capital outwards, while tax monies and resources flowed inward toward the capital. For tax purposes, as Bol notes, it was not necessary to know exact boundaries of different landholdings or territories; it was sufficient to know where the controlling centers were located and to relate these centers to households or landowners that could be held accountable for tax obligations. Accordingly, for his project he uses points and lines rather than polygons such as the USGS quads that form the historical basis for the Western Railroads project. What relationality means, and how it is constituted, are thus historically and culturally specific.

Moreover, in China administrative hierarchies were recorded much earlier than cultural information. The oldest records compiled information primarily for the purposes of taxation and social order; only later were local gazetteers employed by wealthy families to record histories of the places in which the families resided, along with social, economic, and cultural information specific to those communities. As Bol notes, "Gazetteers represented a confluence of interests between local officials and local elite families, those with the ability of educate their children, organize local defense militias, establish water conservancy practices and contribute to schools, bridges, walls and religious institutions" (2008:48). The historical progression from administrative and governmental units to cultural information can be represented through layers, with the older information represented as a network of georeferenced points serving as the base layer on which cultural information is overlaid. The assumption of a universal trajectory of "progress" through time is disrupted by representing historical progression as a succession of layers rather than a linear time line. This configuration shows change through time without implying that the earlier periods were "backward" relative to the later ones, while also showing that the ways in which the layers emerged are historically and culturally specific. Georeferencing to absolute space is here integrated with representations of relationality, drawing into question Massey's assumption of an either/or choice between them.

Space and Temporality

With only a little tongue in cheek, Richard White suggests that "if space is the question, movement is the answer" (2010:para. 17). As he points out, some particularly ingenious mapmakers have found ways to incorporate time into their representations, for example in the French statistician and engineer Charles Joseph Minard's famous map of Napoleon's march into Russia. For a number of reasons, however, continuously changing temporal events have been difficult to incorporate into maps, which historically have focused on spatial relationships rather than temporal events. This is particularly true of paper maps, which are expensive to create and therefore are designed to last for a relatively long time, which in itself tends to exclude features that change over time: a map might show the topographic features of a field, for example, but would not show the cows grazing upon it. With digital maps, the possibility of updates makes temporal events more likely to be included, but even here there are limits. Google Earth images, for example, which many viewers naïvely regard as a "snapshot," are actually created through multiple images stitched together, which sometimes show incongruities when temporal events such as shadows and water ripples are incorporated.

When relational databases are used, time tends to enter as static dates, as is the case with the construction dates referenced in the Western Railroads project, or as time slices, as in the databases showing track lengths at specific dates. Acknowledging these limitations, Michael Goodchild (2008) argues that object-oriented databases have the potential to incorporate temporality in more flexible ways than relational databases. Designed to work with object-oriented programming languages, object-oriented databases store objects that can include complex data types such as video, audio, graphs, and photos. At issue is more than complex versus simple data types, however. In theory, databases are models of the world, and relational and object-oriented databases conceive of the world (or better, assume and instantiate world versions through their procedures) in fundamentally different ways. Relational database representations imply that the world is made up of atomized data elements (or records) stored in separate tables, with similar data types grouped together. As we have seen, these records can be manipulated through declarative statements such as "join," "select," and other commands. The data and commands occupy separate ontological categories, with the data stored in tables and the commands operating upon the tables. One advantage of this world model, as we saw earlier with the Western Railroads project, is that datasets can be extended simply by adding more tables, without needing to revise or change the tables that already exist in a database.

Object-oriented databases, by contrast, divide the world into abstract entities called classes. These classes function somewhat like objects in the real world; they can be manipulated and moved around without needing to know exactly how they are constituted on the inside (a characteristic known as encapsulation). Classes have both attributes (or characteristics) and functionalities (or actions that can be performed upon them). The class "shape," for example, might have functionalities such as changing size, spatial orientation, etc. Classes are arranged hierarchically, and the attributes of the higher class can be inherited by the lower class. For example, the class "shape" might have lower classes such as "circle," "square," etc., and the functionalities associated with "shape" can be inherited by the lower classes without needing to specify them again.

Moreover, relational databases and object-oriented databases operate in distinctive ways through their query languages. In SQL used with relational databases, declarative statements concatenate data records drawn from different tables. Object-oriented databases, on the contrary, work through pointers that reference an object through its location in the database. The object approach is navigational, whereas the relational approach is declarative. In addition, the encapsulation approach of object-oriented databases is fundamentally at odds with the assumption underlying relational databases, which is that data should be accessible to queries based on data content rather than predefined access paths. Object-oriented databases encourage a view of the world that is holistic and behavioral, whereas relational databases see the world as made up of atomized bits that can be manipulated through commands.

These differences notwithstanding, the contrast should not be overstated. In practice, programmers routinely coordinate object-oriented programming languages (now the dominant kind of language) with relational databases (which remain the dominant form of database) through object-relational mapping (ORM).[5] This is a programming technique that converts relational databases into "virtual object databases" that can be manipulated from within object-oriented languages. In effect, the ORM software translates between nonscalar values (for example, a "person-object" that contains name, address, and phone numbers) and scalar values that separate the object into its component elements. These operations are not foolproof, for in mass deletions and joins (i.e., the concatenation of several data elements into a single expression), problems may arise.

Another solution is to use object-oriented databases. Michael Goodchild (2008) argues that object-oriented databases have powerful advantages because they do a better job of representing change over time. Originally ArcGIS, Goodchild points out, was modeled on the metaphor of the print map; it achieved complexity by creating software that functioned like a container holding maps that can be layered on top of one another. To translate this metaphor into relational databases, GIS regards a map "as a collection of *nodes* (the points where lines come together or end), *arcs* (the lines connecting nodes and separating areas), and *faces* (the areas bounded by the arcs)" (182). In this representation, as Goodchild notes, each arc connects to two nodes, and each arc bounds two faces. Therefore, in a relational database, a map can be represented through three tables: arcs, nodes, and faces. The arcs are generally curved and therefore must be represented as multiple line segments, the number varying with the length and curvature of the arc. However, a relational database requires that there be a constant number of entries per row. This means that the table showing the polylines would have to have a number of columns equal to the greatest number of polylines, with the result that most of the table would be empty. As a workaround, polylines and their coordinates were stored outside the relational model, which required developers to maintain two databases, one in tabular form and the other for the polylines.

Since GIS was originally conceived as a container for maps, in earlier decades it suffered from the same inability to represent time that is true of maps. Goodchild notes that "many authors have commented on the inability of 1990s GIS to store information on change through time, information about three-dimensional structures, and information about hierarchical relationships in geographic information" (2008:184). A solution presented itself in object-oriented databases. For a GIS object-oriented database, the most general classes would be the arcs, points, nodes, and polylines that define geographic shapes. More specific classes would be nested under their "parent." For example, in an object-oriented database on military history (the example Goodchild uses), under "points" would be "engagement," "camp," and "headquarters"; under polyline would be "march" and "front."

The movement away from the map metaphor meant that GIS object-oriented databases could "store things that were fundamentally not mappable," Goodchild explains, including classes with "attributes and locations that changed through time" (2008:186). "The only requirement was that the geographic world be conceived as populated by discrete, identifiable, and countable objects in an otherwise empty space" (186). The kinds of relationality that can be represented within a class's functionalities and therefore between classes are more flexible than the kinds of relations that can be represented through declarative commands operating on relational databases, an especially important consideration when temporal events are being represented. For example, in addition to inheritance, objects in an object-oriented database also have the properties of aggregation (the ability for classes to be joined together) and association (the ability to represent flows between classes).

Their advantages notwithstanding, object-oriented databases also have significant limitations. Whereas the standards for SQL are very well established with relational databases, thus allowing for extensive interoperability between databases, the standards for object-oriented databases are more various, including in the areas of connectivity, reporting tools, and backup and recovery standards, making it difficult for a database constructed according to one set of standards to be joined with another database constructed according to different standards. Moreover, whereas relational databases have a well-established formal mathematical foundation, object-oriented databases do not. Partly for these reasons, relational databases remain the standard, with object-oriented databases used mostly in specialized fields such as telecommunications, engineering, high-energy physics, and some kinds of financial services. As long as this is the situation, the open-ended temporality that Massey yearns for will be difficult to represent, and nonspatial attributes are likely to be georeferenced to absolute space as the bottom layer of representation.

The larger point here is the contrast between discursive and database modes of explanation and the ways in which they modify one another. Massey's discursive analysis of "lively" space can inspire database-driven projects to maximize interrelationality through different layers and interactions within and between the data represented in layers. At the same time, the requirements for extensibility and interoperability of database projects indicate that absolute space cannot be dispensed with altogether, as Massey seems to imply. The contrast between relational and object-oriented databases shows a continuing evolution in software that moves in response to the desire of historians to incorporate temporality as movement rather than as static dates. There are many more complications to this story, for example hybrid database systems that combine relational and object-oriented forms and/or port data between them (for a description of Matisse, one such hybrid form, see http://www.15seconds.com/issue/030407.htm). Other possibilities are databases that focus on events rather than places and dates (Mostern and Johnson 2008). In this case, a sense of narrative may be recovered by indicating causal relationships between events. Such a database may be created through relational tables that are extensible and that can be queried to create interactive time lines, spatiotemporal applications, and time lines that represent relationships and narrative. The advantages of an event-focused gazetteer over a GIS or place-name gazetteer include the ability to capture the dynamism of historical interactions, the contexts of spatial organizations, and, perhaps most important, the "interpretive character of historical practice" (Mostern and Johnson 2008:1093).

The contributors to *The Spatial Humanities: GIS and the Future of Humanities Scholarship* (Bodenhamer, Corrigan, and Harris 2010) agree on the limitations of GIS for humanities scholarship, pointing to the tension between the ambiguities of humanities data and approaches and the necessarily quantitative and unambiguous nature of database organizations. They suggest a range of solutions, among which is the provocative idea of a "Pareto GIS" (Harris, Rouse, and Bergeron 2010). Introducing the concept, Harris, Rouse, and Bergeron recall the history of critical GIS, a movement that sought to critique the quantitative and "objective" assumptions of GIS, and the subsequent merger of the two approaches in the field known as "GIS and society." They suggest that the issues in those debates are now being restaged in the spatial history movement. In response to the tensions noted above, especially those between database and narrative, they identify the need for approaches responsive to "the story narrative as presented by the text book which is in contrast to the non-linear digital world of GIS" and "the need to provide a spatial storytelling equivalent" (126). They envision a solution in the "Geospatial Semantic Web, built on a combination of GIS spatial functionality and the emerging technologies of the Semantic Web," which would provide "the core of a humanities GIS able to integrate, synthesize, and display humanities and spatial data through one simple and ubiquitous Web interface" (129).

The Pareto concept mentioned above refers to the 80:20 rule proposed by Italian economist Vilfredo Pareto, whose research indicated that 80 percent of income in Italy went to 20 percent of the people. The 80:20 rule has subsequently been shown to apply to a wide range of phenomena, from wardrobe selection (20 percent of clothes are worn 80 percent of the time) to purchasing patterns (20 percent of customers purchase 80 percent of a given firm's products). Harris, Rouse, and Bergeron propose that "only 20 percent of GIS functionality is sufficient to garner 80 percent of the geospatial humanities benefits" (2010:130). They interpret this to mean that rather than the cumbersome functionality and steep learning curve of a standard GIS data system, 80 percent of its benefits can be gathered from ubiquitous systems such as Google Maps and Google Earth, for which users can provide local information and multimedia content. These layers can include narrative material, which forms a bridge between traditional narrative histories and spatial history capabilities.

Whatever the database forms or data management systems, spatial history projects aim to preserve the interrogative function of history that sees places, dates and events as objects of inquiry as well as analysis. This point is emphasized by Richard White at the conclusion of his essay on spatial history:

> One of the important points that I want to make about visualizations, spatial relations, and spatial history is something that I did not fully understand until I started doing this work and which I have had a hard time communicating fully to my colleagues: visualization and spatial history are not about producing illustrations or maps to communicate things that you have discovered by other means. **It is a means of doing research**; it generates questions that might otherwise go unasked; it reveals historical relations that might otherwise go unnoticed, and it undermines, or substantiates, stories upon which we build our own versions of the past. (2010:para. 36; boldface in original)

To this I would add that spatial history demonstrates the transformative power that digital technologies can exert on a traditionally print-based field, even as the digital methods are also changed by imperatives articulated within print media.

As database-management systems and technologies continue to evolve, the two complementary modes of narrative and database will no doubt find new kinds of rapprochements, including methods of enfolding one into the other. Nevertheless, a crucial difference will likely always separate them as models for understanding the world: database technology relies for its power and ubiquity on the interoperability of databases, whereas narrative is tied to the specificities of individual speakers, complex agencies, and intentions only partially revealed. That is, narratives intrinsically resist the standardization that is the greatest strength of databases and a prime reason why they are arguably becoming the dominant cultural form. This abiding tension, as we will see in the next chapter, provides a complex field of interaction, which contemporary print novels interrogate. Although narratives will not disappear, their forms and functions are being transformed by the seemingly irresistible force of digital databases.

7

Transcendent Data and Transmedia Narrative

Steven Hall's *The Raw Shark Texts*

As chapter 5 argues, database is unlikely to displace narrative as a human way of knowing. Nevertheless, many kinds of professional knowledge structures are moving away from narrative modes into database configurations. A case in point is social work. Analyzing this shift, Nigel Parton (2008) traces the movement from a narrative base for social work, in which "a large amount of knowledge was undocumented and existed primarily in people's heads," to information as a "self-contained substance which can be shared, quantified, accumulated, compared and stored in a database" (262). Knowledge "which cannot be squeezed into the required format disappears or gets lost," he continues, noting that "stories of violence, pain and social deprivation can only be told within the required parameters to the point that they may not be stories at all" (262). This trend in professional knowledge structures is mirrored by the growth of databases in nearly every sector of society.

Even as the data stored in databases has exploded exponentially, the percentage accessible (or indeed even known) to the public has shrunk. The IRS, the US military, and the Department of Homeland Security all maintain huge inaccessible databases, not to mention the large corporate databases of Walmart, Microsoft, and other commercial interests. The creepiness of knowing that one's everyday transactions depend on invisible databases suggests an obvious point of intervention for contemporary novels: subverting the dominance of databases and reasserting the priority of narrative fictions. Steven Hall's remarkable first novel, *The Raw Shark Texts* (S. Hall [2007] 2008a; hereafter *RST*), creates an imaginative world that performs the power of written words and reveals the dangers of database structures. The narrative begins when Eric Sanderson, the protagonist, says, "I was unconscious. I'd stopped breathing" (3). As he fights for life and breath, he realizes that he knows nothing about who he is; all of his memories have mysteriously vanished. The backstory is related through letters he receives from his former self, "The First Eric Sanderson," who tells him he is the victim of a "conceptual shark," the Ludovician, which feeds on human memories and hence the human sense of self. As the Second Eric Sanderson journeys to discover more about his past, the text explores what it would mean to transport a (post)human subjectivity into a database, at the same time that it enacts the performative power of imaginative fiction conveyed through written language. To contextualize his intervention, I will return to issues discussed in chapter 6 about the standardization of data, a crucial requirement for integrating one set of databases with another. Only when the standards are consistent is it possible for databases to link up and expand without limit.

The Emergence of Database Interoperability

Why resist the encroachment of databases on knowledge structures? Aside from issues of surveillance, secrecy, and access, databases also raise serious questions, as Parton suggests (2008), about the kinds of knowledge that are lost because they cannot fit into standardized formats. Alan Liu (2008b) identifies the crucial move in this standardization as "*the separation of content from material instantiation or formal presentation*" (216; emphasis in original). Liu points out that contemporary web pages typically draw from backend databases and then format the information locally with XML tags. This separation of content from presentational form allows for what Liu sees as the imperatives of postindustrial information, namely, that discourse be made transformable, autonomously mobile, and automated. He locates the predecessor technologies in the forms, reports, and gauges of John Hall at Harpers Ferry Armory and the industrial protocols of Fredrick Winslow Taylor. While these early prototypes seem on their faces to be very different from databases and XML, Liu argues that they triggered "the exact social, economic, and technical *need* for databases and XML" (2008b:222; emphasis in original) insofar as they sought to standardize and automate industrial production.

The shift from modern industrialization to postindustrial knowledge work brought about two related changes. First, management became distributed into different functions and automated with the transition from human managers to database software. Second, content became separated not only from presentational form but also from material instantiation. As we have seen, relational databases categorize knowledge through tables, fields, and records. The small bits of information in records are recombinable with other atomized bits through search queries carried out in such languages as SQL. Although database protocols require each atom to be entered in a certain order and with a certain syntax, the concatenated elements can be reformatted locally into a wide variety of different templates, web pages, and aesthetics. This reformatting is possible because when they are atomized, they lose the materiality associated with their original information source (for example, the form a customer fills out recording a change of address). From the atomized bits, a sequence can then be rematerialized in the format and order the designer chooses.

As Liu argues, postindustrial knowledge work, through databases, introduced two innovations that separate it from industrial standardization. By automating management, postindustrial knowledge work created the possibility of "management of, and through, media" (A. Liu 2008b:224), which leads in turn to the management of management, or what Liu calls "metamanagement" (227). The move to metamanagement is accompanied by the development of standards, and governing them, standards of standards. "The insight of postindustrialism," Liu writes, "is that there can be metastandards for making standards" (228). Further, by separating content from both form and materiality, postindustrial knowledge work initiated variable standardization: standardization through databases, and variability through the different interfaces that draw upon the databases (234). This implies that the web designer becomes a "cursor point" drawing on databases that remain out of sight and whose structures may be unknown to him. "Where the author was once presumed to be the originating transmitter of a discourse . . . now the author is in a mediating position as just one among all those other managers looking upstream to previous originating transmitters—database or XML schema designers, software designers, and even clerical information workers (who input data into the database or XML source document)" (235).

Extending the analysis of narrative and database in chapter 5, I will now compare the requirements associated with postindustrialized knowledge work to narrative fiction. Whereas data elements must be atomized for databases to function efficiently, narrative fiction embeds "data elements" (if we can speak of such) in richly contextualized environments of the phrase, sentence, paragraph, section, and fiction as a whole. Each part of this ascending/descending hierarchy depends on all the other parts for its significance and meaning, in feedback and feedforward loops called the hermeneutic circle. Another aspect of contextualization is the speaker (or speakers), so that every narrative utterance is also characterized by viewpoint, personality, and so forth. Moreover, narrative fictions are conveyed through the material instantiations of media, whether print, digital, or audio. Unlike database records, which can be stored in one format and imported into an entirely different milieu without changing their significance, narratives are entirely dependent on the media that carry them. Separated from their native habitats and imported into other media, literary works become, in effect, new compositions, notwithstanding that the words remain the same, for different media offer different affordances, navigational possibilities, etc. (see Hayles 2002 for an elaboration of this point).

The dematerialization and standardization of databases, in contrast to the opposite qualities of embeddedness and causal linkages of narrative, suggest other reasons narrative continues to be essential to the human lifeworld. As Liu's analysis makes clear, databases raise serious questions about the atomization of information, the dispersion of authorship, and the corporatization of contemporary knowledge. These concerns set the stage for my discussion of the strategies by which *RST* instantiates and performs itself as a print fiction—strategies that imply a sharp contrast between narrative and database, especially in light of the standardizations that have made databases into the dominant form of postindustrial knowledge work.

Two Different Kinds of Villains

A major villain in the text, Mycroft Ward, personifies the threat of data structures to narrative fiction. Mycroft Ward,[1] a "gentleman scientist" (S. Hall [2007] 2008a:199) in the early twentieth century, announced that he had decided not to die. His idea was as simple as it was audacious: through a written list of his personality traits and "*the applied arts of mesmerism and suggestion*," (200; emphasis in original), he created "the arrangement," a method whereupon his personality was imprinted on a willing volunteer, a physician named Thomas Quinn, devastated by the loss of his wife.[2] On his deathbed, Ward transferred all his assets to Quinn, who then carried on as the body inhabited by Ward's self. His experiences with World War I impressed upon him that one body was not enough; through accident, warfare, or other unseen events, it could die before its time. Ward therefore set about to standardize the transfer procedure, recruiting other bodies and instituting an updating procedure (held each Saturday) during which each body shared with the others what it had learned and experienced during the week. In addition, in order to strengthen the desire for survival, Ward instituted an increased urge toward self-preservation. This set up a feedback loop such that each time the standardizing process took place, the result was a stronger and stronger urge for survival. In the end, all other human qualities had been erased by the two dominant urges of self-protection and expansion, making Ward a "thing" rather than a person. By the 1990s, "the Ward-thing had become a huge online database of self with dozens of permanently connected node bodies protecting against system damage and outside attack. The mind itself was now a gigantic over-thing, too massive for any one head to contain, managing its various bodies online with standardizing downloads and information-gathering uploads. One of the Ward-thing's thousands of research projects developed software capable of targeting suitable individuals and imposing 'the arrangement' via the Internet" (204).

The above summary takes the form of a nested narrative, one of many that pepper the main narrative. This one is narrated by Scout, the Second Eric's companion, herself a victim of an Internet attack by the "Ward-thing" to take over her body. Saved only because her younger sister pulled out the modem cord to make a phone call, Scout is aware that some small part of her is inhabited by the Ward-thing, while the online database contains some small part of her personality. She is alive as an autonomous person because she fled her previous life and now lives underground in the "unspace" of basements, cellars, warehouses, and other uninhabited and abandoned places that exist alongside but apart from normal locales. The precariousness of her situation underscores the advice that the First Eric Sanderson gives to his successor: "There is *no* safe procedure for [handling] electronic information" (81; emphasis in original).

The qualities Liu identifies as the requirements to make postindustrial information as efficient and flexible as possible—namely transformability, autonomous mobility, and automation—here take the ominous form of a posthuman "thing" that needs only find a way to render the standardizing process more efficient to expand without limit, until there are a million or billion bodies inhabited by Mycroft Ward, indeed until the world consists of nothing but Mycroft Ward, a giant database hidden from sight and protected from any possible attack.[3] Ward in this sense becomes the ultimate transcendental signified, which Liu (following Derrida) identifies as "content that is both the center of discourse and—precisely because of its status as essence—outside the normal play or (as we now say) networking of discourse" (A. Liu 2008b:217). At that terminal point, there is no need for stories, their function having been replaced by uploading and downloading standardized data.

Whereas traditional narrative excels in creating a "voice" the reader can internalize, with the advent of databases the author no longer crafts language so that it seems to emanate from the words on the page. Rather, the author function is reduced to creating the design and parameters that give form to content drawn from the databases. As Liu points out, the situation is even worse than reducing the author function to a "cursor position," for the "origin of transmission in discourse network 2000 is not at the cursor position of the author. Indeed, the heart of the problem of authorship in the age of networked reproduction is that there is no cursor point" (A. Liu 2008b:235–36), only managers downstream of invisible databases that dictate content. This threat to the author's subjectivity (and implicitly to the reader's) is dramatically apparent in the significantly named Mr Nobody,[4] one of Ward's node bodies. When the Second Eric is invited to meet with Nobody, he first finds a confident, well-dressed man working on a laptop. As the conversation proceeds, however, Nobody becomes increasingly disheveled and drenched in sweat until "liquid streamed off him into small brown pools around the legs of his chair," (S. Hall [2007] 2008a:143), a description that reinforces the role water and liquidity play in the text. At one point he turns away from Eric and begins conversing with an invisible interlocutor: "It's too long, the weave has all come apart—loose threads and holes, *he's* showing through, you know how it gets just before the pills" (143; emphasis in original). Scout later holds up the confiscated pill bottles, marked with such labels as "CONCENTRATION," "REASONING," "STYLE." "This is him," she comments. "The closest thing to a him there was anyway. This is what was driving the body around instead of a real self" (178). She observes,

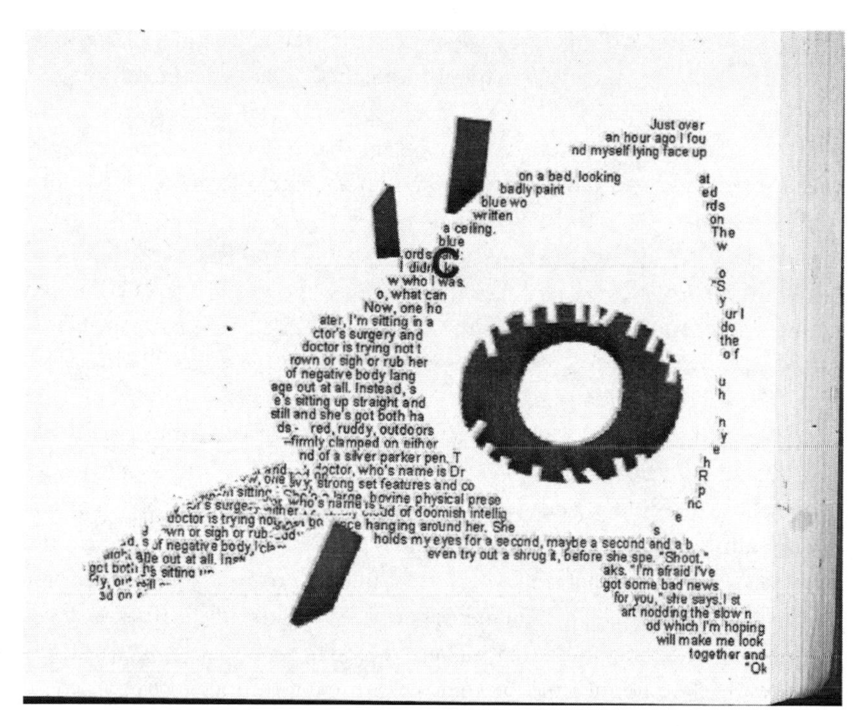

Figure 7.1 Graphic representing Ludovician shark as composed of words. From *The Raw Shark Texts*, created and owned by Steven Hall. First published in 2007 by Canongate Books. Reproduced with permission.

"He wasn't really a human being anymore, just the idea of one. A concept wrapped in skin and chemicals" (178).

Whereas Ward represents the complete separation of form and content (nude bodies versus online database), the text's other major villain, the Ludovician shark, embodies the complete fusion of form and content. Graphic and verbal representations of the "conceptual shark" depict it as formed through the collapse of signifier and signified into one another. Shown as a graphic, the shark's body is composed of words and typographical symbols (fig. 7.1). Moreover, the text that forms the body changes from image to image, making good the premise that the shark's flesh forms from the memories he has ingested (in fig. 7.1, some of the text seems to refer to memories that could have belonged to the First Eric).

Represented verbally, the shark combines the morphology of an actual shark with flesh made up of ideas, language, and concepts, as shown in this passage when Eric and Scout are being pursued by the Ludovician: "Less than fifty yards behind us and keeping pace, ideas, thoughts, fragments, story shards, dreams, memories were blasting free of the grass in a high-speed spray. As I watched, the spray intensified. The concept of the grass itself began to lift and bow wave into a long tumbling V. At the crest of the wave, something was coming up through the foam—a curved and rising signifier, a perfectly evolved fin idea" (158). Such descriptions create homologies between action in the diegesis, the materiality of the shark as it appears within the diegesis, and the materiality of the marks on the page.

A further transformation occurs when not just the shark but the surrounding environment collapses into conceptual space that is at once composed of words and represented through words on the page. Illustrative is the striking scene when the Second Eric is attacked by the shark in his own home, and suddenly the mundane architecture of floor and room gives way to a conceptual ocean. "The idea of the floor, the carpet, the concept, feel, shape of the words in my head all broke apart on impact with a splash of sensations and textures and pattern memories and letters and phonetic sounds spraying out from my splashdown. . . . I came up coughing, gasping for air, the idea of air. A vague physical memory of the actuality of the floor survived but now I was bobbing and floating and trying to tread water in the idea of the floor, in fluid liquid concept, in its endless cold rolling waves of association and history" (59).

The Dangerous Delights of Immersive Fiction

Jessica Pressman (2009) rightly notes that the entanglement of print mark with diegetic representation insists on the work's "bookishness," a strategy she suggests is typical of certain contemporary novels written after 2000 in reaction to the proclaimed death of print. Here fear rather than necessity is the mother of invention; she proposes writers turn to experimental fiction to fight for their lives, or at least the life of the novel. Discussing the Ludovician shark, she likens its devouring of memories to a fear of data loss (and to Alzheimer's). In addition to these suggestions, I think something else is going on as well, a pattern that emerges from the way depth is handled in the text. The shark is consistently associated with a form that emerges from the depths; depth is also associated with immersion, in Eric's "splashdown" in his living room, for example, and his later dunking when he goes hunting for the shark.

This latter episode, which forms the narrative's climax, follows the plot of *Jaws* almost exactly. As in *Jaws*, the climax is nonstop heart-pounding action, arguably the clearest example in the text of what is often called "immersive fiction" (see, for example, Mangen 2009). The link between the acts of imagination that enable a reader to construct an English drawing room or a raging ocean from words on the page is, in mise en abyme, played out for us within the very fiction upon which our imaginations operate. Scout and Trey Fidorous, the man who will be the guide to locate and attack the Ludovician, put together the "concept" of a shark-hunting boat, initially seen as a eccentric conglomeration of planks, speakers, an old printer, and other odds and ends laid out in the rough shape of a boat. "It's just stuff," Fidorous comments, "just beautiful ordinary things. But the idea these things embody, the meaning we've assigned to them in putting them together like this, that's what's important" (S. Hall [2007] 2008a:300). While Scout and Fidorous create the assemblage, Eric's task is to drink and be refreshed by a glass full of paper shards, each with the word "water" written on them. At first he has no idea how to proceed, but emerging from a dream in which he falls into shark-infested waters, he suddenly senses that his pant leg is wet—moistened by water spilled from the overturned glass. With that, torrents of water flood into the space, and the flat outline of the boat transforms into an actual craft floating on the ocean. Rubbing "a hand against a very ordinary and very real railing," Fidorous tells him that the boat "is the current collective idea of what a shark-hunting boat should be" (315). Hovering between believing in the scene's actuality and seeing it as the *idea* of the scene, Eric is advised by Fidorous, "It's easier if you just accept it" (317). With that, the scene proceeds in immersive fashion through the climax.

What is the connection between immersive fiction and immersion in dangerous waters inhabited by a memory-eating shark? At the same time the text defends the aesthetic of bookishness, it also presents fiction as a dangerous activity to be approached with extreme care. In the one case, we see the strongly positive valuation the author places on the print book; in the other case, the valuation is more ambivalent or perhaps even negative. Nevertheless, the two align in attributing to fiction immense powers to bring a world into being. The most dramatic testimony to fiction's power comes when the Second Eric finds the courage to enter the upstairs locked room in the house he inhabits/inherits from his predecessor. Inside is a red filing cabinet, and in the top drawer, a single red folder with a typed sheet that begins "Imagine you're in a rowing boat on a lake," followed by evocative details that are the stock in trade of immersive narratives. Then the narrative commands,

> Stop imagining. Here's the real game. Here's what's obvious and wonderful and terrible all at the same time: the lake in my head, the lake I was imagining, has just become the lake in your head. . . . There is some kind of flow. A purely conceptual stream with no mass or weight or matter and no ties to gravity or time, a stream that can only be seen if you choose to look at it from the precise angle we are looking from now. . . . Try to visualize all the streams of human interaction, of communication. All those linking streams flowing in and between people, through text, pictures, spoken words and TV commentaries, streams through shared memories, casual relations, witnessed events, touching pasts and futures. . . . This waterway paradise of all information and identities and societies and selves. (54–55)

We know from Fidorous that conceptual fish, including the Ludovician, evolved in the information streams that have, during the twentieth and twenty-first centuries, reached unprecedented density and pervasiveness (although the Ludovician is an ancient species).

The passage continues with an ominous evocation of the shark's appearance: "Know the lake; know the place for what it is . . . take a look over the boat's side. The water is clear and deep. . . . Be very, very still. They say life is tenacious. They say given half a chance, or less, life will grow and exist and evolve anywhere. . . . Keep looking and keep watching" (55). The lake, then, is a metonym for both fiction and the shark, a conjunction that implies immersive narratives are far from innocent. It is no accident that Eric, after reading the text quoted above, experiences the first attack by the Ludovician in his living room.

Why would fiction be "terrible" as well as wonderful, and what are the deeper implications of double-valued immersion? Like the shark, immersive fictions can suck the reader in. In extreme cases, the reader may live so intensely within imagined worlds that fiction comes to seem more real than mundane reality—as the characters Madame Bovary and Don Quixote testify. A related danger (or benefit?) is the power of fiction to destabilize consensus reality, as in the narrator's pronouncement in *Gravity's Rainbow* that "It's all theatre" (Pynchon 1974:3). Once certain powerful fictions are read, one's mental landscape may be forever changed (this happened to me when I first encountered Borges's fictions). Recall Henri Lefebvre's proclamation ([1974] 1992) that social practices construct social spaces; narrative fiction is a discursive practice that at once represents and participates in the construction of social spaces. Through its poetic language, it forges relationships between the syntactic/grammatical/semantic structures of language and the creation of a social space in which humans move and live, a power that may be used for good or ill. No wonder the First Eric says of his evocation of a fictional lake, quoted above, that this text is "'live' and extremely dangerous" (Hall [2007] 2008a:69).

Countering the dangers of immersive fiction is the prophylaxis of decoding. To imagine a scene, a reader must first grasp the text through the decoding process of making letters into words, words into narratives. Reading an immersive fiction, the fluent reader typically feels that the page has disappeared or become transparent, replaced by the scene of her imagination. In *RST*, this process is interrupted and brought into visibility through elaborate encoding protocols, notably with the First Eric's flashback narratives. The first such narrative arrives in a package that includes a video of a lightbulb flickering on and off in a dark room, accompanied by two notebooks, one full of scribbles, the other a handwritten text. The letter explaining the package says that the on-off sequences of the flickering bulb send a message in Morse code, the letters of which are transcribed into one of the notebooks. The decoding does not stop there, however, for the Morse code letters are further encrypted through the QWERTY code. Locate each Morse code letter on a standard typewriter keyboard, the letter instructs; note that it is surrounded by eight other letters (letters at the margins are "rolled around" so they are still seen as surrounded by eight others). One of the eight is then the correct letter for a decrypted plaintext, with the choice to be decided by context (fig. 7.2).

This decryption method is tricky, for it means that one needs a whole word before deciding if a given letter choice is correct. Not coincidentally, it also emphasizes the dependence of narrative on context and sequential order. Whole paragraphs must be decrypted before the words they comprise may be considered correct, and the entire narrative before each of the paragraphs can be considered stable. The reader is thus positioned as an active decoder rather than a passive consumer, emphasized by the hundreds of hours required to decrypt the "Light Bulb Fragments." These textual sections, purporting to be the plaintexts of the encodings, recount the adventures of the First Eric with his girlfriend Clio Ames on the Greek island of Naxos.

The decoding does not end here, however, for Fidorous points out to Eric that encoded on top of the QWERTY code is yet another text, encrypted through the angular letters formed by the vector arrows that emerge from the direction the decoding took from one QWERTY keyboard letter to another (see fig. 7.3). This text, entitled "The Light Bulb Fragment (Part Three/

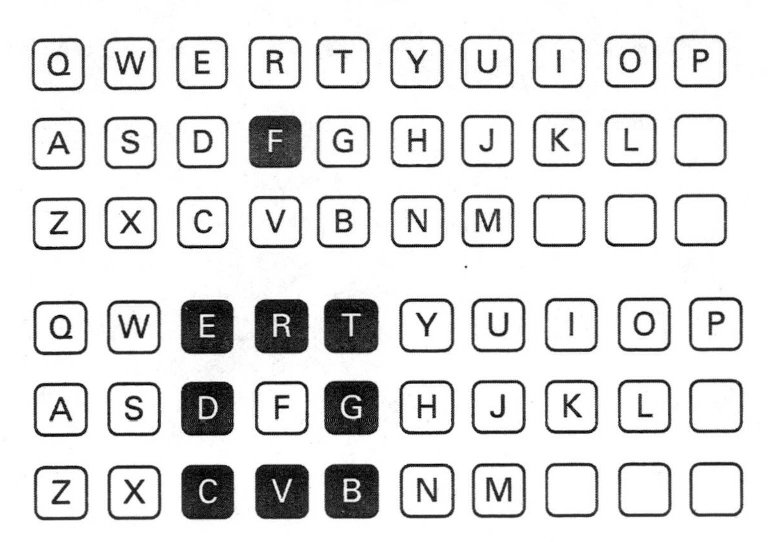

Figure 7.2 QWERTY decryption method. From *The Raw Shark Texts*, created and owned by Steven Hall. First published in 2007 by Canongate Books. Reproduced with permission.

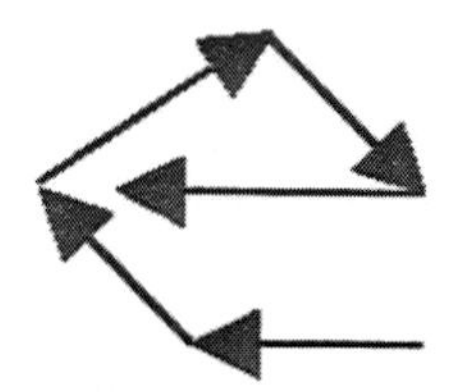

Figure 7.3 Double encryption through vectors formed from QWERTY letters. From *The Raw Shark Texts*, created and owned by Steven Hall. First published in 2007 by Canongate Books. Reproduced with permission.

Encoded Section)," begins, "Everything is over" (S. Hall [2007] 2008a:410), with the First Eric narrating the denouement following Clio's death while scuba diving. Since the double encryption begins with the very first series of letters in the "Light Bulb Fragment," the tragic end of the First Eric's time with Clio is written simultaneously with the beginning, a fantastically complex method of double writing that pays tribute to the complex temporalities for which narratives are justly famous.

The encodings within encodings represent not only protection against immersion but protection against the Ludovician as well. Writing serves two opposite functions in the text. On the one hand, it is part of the stream of information from which the Ludovician evolved and through which it catches the scent of Eric Sanderson (First and Second) as the Ludovician pursues him. On the other hand, as fiction, it also can be used to disguise Eric's trail and protect against the Ludovician, as can other written documents such as letters, library books, and so on. The First Eric explains: "Fiction books also generate illusionary flows of people and events and things that have never been, or maybe have only half-been from a certain point of view. The result is a labyrinth of glass and mirrors which can trap an unwary fish for a great deal of time" (68). Nested narratives, because of their complexity, are especially effective. The First Eric says he has an "old note" which "says that some of the great and more complicated stories like the *Thousand and One Nights* are very old protection puzzles, or even idea nets by which ancient people would fish for and catch the smaller conceptual fish" (68). In addition to encoding and encryption, also effective as protection are material objects associated with writing or covered by writing. When Scout and Eric are pursued by the Ludovician coming up through the grass, she gives Eric a "bomb" to throw composed of old typewriter keys and other such metal shards. The resulting delay in the shark's pursuit enables them to reach "unspace."

If social spaces are formed through social practices, as Lefebvre asserted, the social practice that both creates and names unspace is writing. The deeper Scout and the Second Eric penetrate into it, the more books, telephone directories, and writing of all sorts emerge as the built environment providing protective camouflage. Their goal is Fidorous's lair, the entrance to which is constructed as a tunnel in which "everything, everything had been covered in words, words in so many languages. Scraps flapped against me or passed under my crawling wrists covered in what I vaguely recognized as French and German, in hard Greek and Russian letters, in old fashioned English with the long 'f's instead of 's's and in Chinese and Japanese picture symbols" (225). Indeed, even the map through the tunnel comes in the form of a word, "ThERa," a Greek island in the Cyclades chain, which includes Naxos (fig. 7.4). As Scout and the Second Eric wend their way through the tunnel, they "read" the word not through linguistic decoding but through embodied movement, feeling in their knees and hands the turns that take them from the top of the "T" to the loop of the "a."

Their embodied transversal emphasizes that writing/reading is not only a semiotic encoding/decoding but a material practice as well. In contrast to the dematerialization that occurs when data are separated from their original instantiations, entered as database records, and reinstantiated as

Figure 7.4 "ThERa" as embodied map. From *The Raw Shark Texts*, created and owned by Steven Hall. First published in 2007 by Canongate Books. Reproduced with permission.

electronic bits, inscriptions on paper maintain a visible, even *gritty* materiality capable of interacting directly with human bodies.

The Raw Shark Texts as a Distributed Media System

However much it celebrates "bookishness," *RST* cannot avoid its embeddedness in a moment when electronic databases have become the dominant cultural form. Along with strategies of resistance, the novel also slyly appropriates other media, including digital media, for its own purposes, instantiating itself as a distributed textual system across a range of media platforms. At the center of the system is the published novel, but other fragments are scattered across diverse media, including short texts at various Internet sites, translations of the book in other languages, and occasional physical sites. The fragments that have been identified so far are listed at The Red Cabinet (S. Hall 2008b). The effort involved in identifying the fragments changes the typical assumption that a reader peruses a novel in solitude. Here the quest for understanding the text, even on the literal level of assembling all its pieces, requires the resources of crowd sourcing and collaborative contributions.

Elsewhere on this forum site, under the general rubric of "Crypto-Forensics," Hall notes that "for each chapter in *The Raw Shark Texts* there is, or will be, an un-chapter, a negative. If you look carefully at the novel you might be able to figure out why the un-chapters are called negatives" (2007b). This broad hint may direct us to the text's climax, when the action becomes most immersive. As the narrative approaches the climax, an inversion takes place that illustrates the power of fiction to imagine a spatial practice undreamed by geographers: space turns inside out. The text's trajectory from the mundane ordinary space in which the Second Eric awakes, the actual locations in England to which he journeys,[5] the unspaces of underground tunnels and protective architecture of written materials he discovers, and finally the imaginative space of the mise en abyme of the shark boat and the ocean on which it floats, may in retrospect be seen as an arc that moves the novel's inscribed spaces further and further from consensus reality. The spatial inversion that occurs at the climax is the fitting terminal point of this arc.

The inversion comes when Eric, carrying a postcard showing the island Naxos, discovers that the original image has been mysteriously replaced by a picture of his house. When he experimentally tries thrusting his hand inside the picture, it starts to come to life, with cars whizzing and birds flying by. He realizes he has an opportunity to return home via this route but refuses to go back to a safe but mundane reality. With that, the picture changes again, to the brightly colored tropical fish the First Eric retrieved from Clio's underwater camera after she died and then had thrown down a dark hole in a foundation. "Something huge happening here. Something so, so important" (S. Hall [2007] 2008a:422), the Second Eric thinks. Earlier when he had a similar sensation, he had thought about a coin that has lain in the dirt, one side up, the other down for years; then something happens, and it gets reversed. "The view becomes the reflection and the reflection the view," he muses (392).

This sets the scene for our understanding of the "negatives." They represent alternative worlds in which the reflection—the secondary manifestation of the view—becomes primary, while the view becomes the secondary reflection. The point is made clear in the negative to chapter 8. Found in the Brazilian edition and available only in Portuguese, it presents as letter 175 from the First Eric. Translated by one contributor to the collective project and revised by another, the letter says in part

> I have dark areas on my mind. Injuries and holes whose bottom I can't see, no matter how hard I try. The holes are everywhere and when you think about them, you can't stop thinking about them. Some holes are dark wells containing only echoes, while others contain dark water in their depths. Inside them I can see a distant full moon, and the silhouette of a person looking back at me. The shape terrorizes me. Is it me down below? Is it you? Maybe it isn't anyone. The view becomes a reflection and . . . something more, something else. (S. Hall 2007c)

Evocative of the lake scene in which the dark water becomes the signifier for the emergence of the Ludovician, the phrase that the First Eric cannot quite complete is fully present in the Second Eric's thoughts in the novel: "The view becomes the reflection, and the reflection the view" (S. Hall [2007]

2008a:392). Simply put, the negatives represent an alternative epistemology that from their perspective is the view, but from the perspective of the "main" narrative represents the reflection.

A cosmological analogy may be useful: the negatives instantiate a universe of antimatter connected to the matter of the novel's universe through wormholes. The parallel with cosmology is not altogether fanciful, for in the "undexes" (indexes that reference the semantic content of the negatives), one of the entries is "Dr Tegmark" (S. Hall 2009), a name that repeats at the novel's end in the phrase "Goodby Dr Tegmark" (S. Hall [2007] 2008a:426). The reference is to Max Tegmark, an MIT physicist, who has written several popular articles about parallel universes. On his website entitled "The Universes of Max Tegmark," the scientist comments, "Every time I've written ten mainstream papers, I allow myself to indulge in writing one wacky one. . . . This is because I have a burning curiosity about the ultimate nature of reality; indeed, this is why I went into physics in the first place. So far, I've learned one thing in this quest that I'm really sure of: whatever the ultimate nature of reality may turn out to be, it's completely different from how it seems." (Tegmark n.d.). The speculative nature of Tegmark's "wacky" articles and his intuition that consensus reality is not an accurate picture is represented in the text through the wormhole that opens in the print text when Eric puts his hand in the photograph. The fact that mundane reality is now the "reflection" (relative to his viewpoint) implies that he is now in an alternative universe, one that is somehow definitively different from the place where he began.

What are the differences, and how are they represented within the text? We can trace the emergence of this alternative universe through the Second Eric's journey into unspace. In addition to wending deeper into the labyrinths of written material that lead to Fidorous, Eric experiences a growing attraction to Scout and an intuition that somehow Scout is not only like Clio but, in a "ridiculous" thought (S. Hall [2007] 2008a:303) that won't leave him alone, actually *is* Clio mysteriously come back from the dead. This, the deepest desire of the First Eric, is why he let the Ludovician shark loose in the first place. The Second Eric discovers in the bedroom that the First Eric occupied in Fidorous's labyrinth a crucial clue; he finds an *Encyclopedia of Unusual Fish*, containing a section entitled "The Fish of Mind, Word and Invention" (263–64). The section refers to an "ancient Native American belief that all memories, events, and identities consumed by one of the *great dream fishes* would somehow be reconstructed and eternally present within it" (265; emphasis in original). Shamans allowed themselves to be eaten by the shark, believing that in this fashion "they would join their ancestors and memory-families in eternal *vision-worlds* recreated from generations of shared knowledge and experience" (265; emphasis in original). Reading about this practice, the First Eric evidently resolved to seek Clio in a similar fashion, a venture that implies the existence of a vision-world in which they could be reunited and live together.

Negative 1 is the "Aquarium Fragment," which presents as the prologue to *RST* (S. Hall 2007a). Alluded to in chapter 7 (S. Hall [2007] 2008a:63), it narrates the backstory to the First Eric's quest. An extra page inserted into the Canadian edition contains the fragment's cover image and title page describing it as "a limited edition lost document from THE RAW SHARK TEXTS by Steven Hall." In pages numbered –21 to –6 (since this is a negative, the numbers would logically go from the more negative to the less negative as the narrative moves forward), the First Eric recounts entering the aquarium, an underground unspace that contains row after row of neglected fish tanks. He finds the tank containing the Ludovician, "the last one in captivity," Jones, his guide, tells him. In front of the tank sit three old men, each looking at the man on his left while furiously scribbling on a pad of paper. Finding one of the pages, Eric reads a description of a man's face in minute detail, recounted in one long run-on sentence that, Jones tells him, runs through all the files in the room. "Direct observation. Always in the present tense, no simile, analogy, nothing remembered or imagined. No full stops. Just only and exactly what they see when looking at the man to their left, in a continuous and ongoing text" (S. Hall 2007a:–8). Eric understands that this writing functions as a "non-divergent conceptual loop," a kind of verbal feedback system that constitutes the shark's cage. Significantly, this is a form of writing that abjures any fictionalizing, any act of the imagination. Walking into the space, Eric experiences a "horrific clarity" that "came into the world, a sense of all things being exactly what they were. . . . I turned slowly on my heels; all three men were staring straight at me. All things filled with relevance, obviousness and a bright four-dimensional truth" (–8).

It is in this state that Eric sees the Ludovician, "partly with my eyes, or with my mind's eye. And partly heard, remembered as sounds and words in shape form. Concepts, ideas, glimpses of other lives or writings or feelings" (–7). There follows a graphic of the kind seen in the main text, of a shark made up of words, which Eric perceives as "swimming hard upstream against the panicking fast flow of my thoughts" (–7). After another, larger graphic (because the shark is getting closer), there follows the sentence "The Ludovician, in my life in every way possible," followed by the upside-down and inverted sentence "My life, into the Ludovician in every way possible" (−6). The semantic mirror inversion of the sentence across the horizontal plane and visual mirror inversion across the vertical axis function as the wormhole that allows the reader to traverse from the alternative universe of the negative back to the "main" narrative of the print novel.

The "negative" prologue is essential for understanding the logic behind the Second Eric's intuition that Scout *is* Clio, as well as for the denouement in which Eric and Scout swim toward Naxos, presumably to reenter the idyllic life the First Eric shared there with Clio. This is, of course, an impossible desire—to bring someone back from the dead and enjoy once again the life that one had before the beloved's death. As an impossible desire, this wish is deeply tied up with the Ludovician and what the "conceptual shark" represents. As we have seen, the Ludovician is constituted through a feedback loop between signifier and signified, mark and concept. As theorized by Saussure, the mark and concept are two sides of the same coin, two inseparable although theoretically distinct entities (Saussure [1916] 1983:649ff.). In rewriting Saussure, Jacques Lacan changed the emphasis so that the concept (or signified) faded into insignificance and the major focus was on the play of signifiers in infinite chains of deferrals, a process arising from what Lacan (1966) described as "the incessant sliding of the signified under the signifier" (419). Lacan theorized that the chain of signifiers was linked with the deferral of socially prohibited desire, specifically desire for the mother. Because desire, to exist as such, implies lack, the chain of signifiers is driven by a logic that substitutes a secondary object for the unattainable object of desire (*le objet petit a*), a substitution displaced in turn by another even more distanced object of desire and so on in an endless process of deferrals and substitutions. Language as a system instantiates this logic in the play of signifiers that can never be definitively anchored in particular signifiers, that is, a system of "floating signifiers" (Mahlman 1970).

What if language, instead of sliding along a chain of signifiers, were able to create a feedback loop of continuous reciprocal causality such that the mark and concept coconstituted each other? Such a dynamic would differ from Saussure, because there is would be no theoretical distance between mark and concept; it would also differ from Lacan, because the signified not only reenters the picture but is inextricably entwined with the signifier. Defying Lacan's logic of displacement, the result might be to enable an impossible desire to be realized, albeit at a terrible cost, since it is precisely on the prohibition of desire that, according to psychoanalytic theory, civilization and subjectivity are built. The Ludovician represents just this opening and this cost: Clio is restored in Scout, but at the cost of annihilating the First Eric's memories and therefore his selfhood. When the wormhole opens for the Second Eric, he has the choice of reentering the universe in which impossible desire can be satisfied or returning to his quotidian life, in which case the Scout-Clio identification would surely disappear, if not Scout herself. When he decides to remain in the universe of the reflection rather than the view, he traverses the wormhole and comes out on the other side.

The mutual destruction of the Ludovician and Mycroft Ward occurs when the two inverse signifiers (the complete separation of form and content in Ward and their complete fusion in the Ludovician) come into contact and cancel each other out in a terrific explosion, an event the text repeatedly compares to matter and antimatter colliding (S. Hall 2008a:246, 318). With that, the wormhole closes forever, and the cost of Eric's choice is played out in the denouement. In a move that could never be mapped through database and GIS technology, the narrative inscribes a double ending.

The first is a newspaper clipping announcing that the dead body of Eric Sanderson has been recovered from foundation works in Manchester. Allusions to the Second Eric's journey echo in the clipping. For example, the psychiatrist making the announcement is named "Dr Ryan Mitchell," the same name Eric encountered in the "Ryan Mitchell mantra" used to ward off the first attack of the Ludovician. The repetition of names implies that his journey through unspace has been the hallucinogenic rambling of a psychotropic fugue (the condition with which Eric's psychiatrist, Dr Randle, had diagnosed him). Extensive parallels support this reading, suggesting that Eric has taken details from his actual life and incorporated them into his hallucination. In this version the distinction between the First and Second Eric is understood as an unconscious defense mechanism to insulate Eric from the pain of Clio's loss. For example, Eric reveals in the triple-encoded section recounting the aftermath of Clio's death that his anguish is exacerbated by late-night phone calls from Clio's father blaming him for her death, with the phone rings represented as "burr burr, burr burr" (S. Hall [2007] 2008a:412). In the shark-hunting episode, the shark's attack on the boat is also preceded by "burr, burr" (418), presumably the sound of it rubbing against the boat's underside. Similarly, the foundation in which Eric's body is found corresponds with the "deep dark shaft" (413) in which the First Eric threw Clio's fish pictures. The image is repeated in the (hallucinated) dark hole into which (the First) Eric descended to find the Ludovician, and the "dark wells" that (the First) Eric perceives in his memory in Negative 8. This reading is further supported by a clever narrative loop in which (the Second)

Eric starts writing his life's story with a magical paintbrush given him by Fidorous, beginning with "*I was unconscious. I'd stopped breathing*" (286; emphasis in original), a move that locates the entirety of (the Second) Eric's narration within his hallucination. Moreover, the close parallels of the shark hunt with *Jaws* (if not a lack of imagination on the author's part) may signal that Eric is appropriating this well-known film and recasting it with himself in the starring role, another indication that the action is happening only in his own hallucination. Finally the tunnel leading into Fidorous's lair in the shape of "ThERa" inscribes not only a Greek island in the Naxos vicinity but also the first letters of the title *The Raw Shark Texts*, forming a recursive loop between Eric's action within the narrative and the narrative as a whole, thus placing the entire story within the confines of a hallucination.

The other version takes the Second Eric's narrative as "really" happening. This reading is reinforced by a postcard Dr Randle receives a week after Eric's body is found announcing that "I'm well and I'm happy, but I'm never coming back," signed "Eric Sanderson" (427). The card's obverse, shown on the next page (428), depicts a frame from *Casablanca* in which the lovers are toasting one another, a moment that in this text is literally the last word, as if to contravene in the film and insist that happy endings are possible after all. Also reinforcing this reading is the reader's investment in the Second Eric's narrative. To the extent a reader finds it compelling, she will tend to believe that the Second Eric is an authentic subjectivity in his own right.

The undecidability of the double ending makes good the eponymous pun on "Rorschach tests." Like an inkblot with two symmetric forms mirroring each other across a vertical axis, the double ending inscribes an ambiguity so deep and pervasive that only a reader's projections can give it a final shape. The ambiguity highlights another way in which narrative differs from database: its ability to remain poised between two alternatives without needing to resolve the ambiguity. As we have seen, in the fields of relational databases, undetermined or unknown (null) values are anathema, for when a null value is concatenated with others, it renders them null as well. In narrative, by contrast, undecidables enrich the text's ambiguities and make the reading experience more compelling.

Narrative is undoubtedly a much older cultural form than database (especially if we consider databases as written forms) and much, much older than the digital databases characteristic of the information age. In evolutionary terms, narrative codeveloped with language, which in turn codeveloped with human intelligence (Deacon 1998; Ambrose 2001). Narrative, language, and the human brain are coadapted to one another. The requirements for the spread of postindustrial knowledge that Alan Liu identifies—transformability, autonomous mobility, and automation—point to the emergence of machine intelligence and its growing importance in postindustrial knowledge work. It is not human cognition as such that requires these attributes but rather machines that communicate with other machines (as well as humans). In contrast to databases, which are well adapted to digital media, narratives are notoriously difficult to generate using intelligent machines, and those that are created circulate more because they are unintentionally funny than because they are compelling stories (see Wardrip-Fruin [2008] for examples). To that extent, narrative remains a uniquely human capacity.

In the twin perils of the Ludovician shark and Mycroft Ward, we may perhaps read the contemporary dilemma that people in developed countries face nowadays. In evolutionary terms, the shark is an ancient life form, virtually unchanged during the millennia in which primates were evolving into *Homo sapiens*. In contrast, the Ward-thing is a twentieth- and twenty-first-century phenomenon, entirely dependent on intelligent machines for its existence. Walking around with Pleistocene brains but increasingly immersed in intelligent environments in which most of the traffic goes between machines rather than between machines and humans, contemporary subjects are caught between their biological inheritance and their technological hybridity. *RST* imagines a (post)human future in which writing, language, books, and narratives remain crucially important. With the passing of the Age of Print, books and other written documents are relieved of the burden of being the default medium of communication for twentieth- and twenty-first-century societies. Now they can kick up their heels and rejoice in what they, and they alone among the panoply of contemporary media, do uniquely well: tell stories in which writing is not just the medium of communication but the material basis for a future in which humans, as ancient as their biology and as contemporary as their technology, can find a home.

8

Mapping Time, Charting Data

The Spatial Aesthetic of Mark Z. Danielewski's *Only Revolutions*

As we saw in chapter 6, databases emphasize spatial displays, whereas narrative embodies complex temporalities. In responding to the overwhelming amounts of data inundating developed societies in the contemporary era (see Dannenberg [2008], Francese [1979], and Johnston [1998] for other literary texts that also respond), Mark Z. Danielewski has launched a bold experimental novel, *Only Revolutions* (2007b; hereafter *OR*), that interrogates the datasphere by accentuating and expanding the role of spatiality in a literary text. In this sense, it displays the effects of data not only at the diegetic level of the narrative but also in the material form of the print codex itself. Among the transformations and deformations the text implements is a profound shift from narrative as a temporal trajectory to a topographic plane upon which a wide variety of interactions and permutations are staged. Whereas narrative temporality proceeds along a one-dimensional line whose unfolding, backtracking, and foreshadowing is carried out through reading practices that typically follow line upon line, a plane has two dimensions through which interactions can take place. Stacking the two-dimensional planes adds a third dimension of depth. In *OR*, the rich dimensionality created by this topographic turn is correlated with an explosive increase in the kinds of reading practices afforded by the text. The results are hybridizations of narrative with data, temporality with spatiality, and personal myth with collective national identity.

Spatial Form and Information Multiplicity

To evaluate this topographical turn, I return to 1991, when Joseph Frank revisited his seminal 1945 essay in "Spatial Form: Some Further Reflections." He rehearses the well-known semiotic model in which *paradigmatic* indicates alternative word choices that define a given term through their differential relations with it, while *syntagmatic* refers to the temporality of syntactic sequence. Envisioned as two perpendicular axes, the model effectively converts a temporal line into a plane of interaction. Since the paradigmatic works together with the syntagmatic, the framework implies spatiality is present in some degree in all literature. Quoting Gérard Genette, he notes that "Saussure and his continuators have brought to the foreground a mode of language that one must call spatial, although we are dealing here, as Blanchot has written, with a spatiality 'whose originality cannot be grasped in terms either of geometrical space or the space of practical life'" (Frank 1991: 124). Whereas Frank had earlier turned up his nose at concrete poetry, he agrees with Genette on the "so-called visual resources of script and topographical arrangement; and of the existence of the Book as a kind of total object" (128). Again from Genette, he focuses on a passage in which the book's materiality comes almost into view: "To read as it is necessary to read [Proust] . . . is really to reread; it is already to have reread, to have traversed a book tirelessly in all directions, in all its dimensions. One may say, then, that the space of a book, like that of a page, is not passively subject to the time of linear reading; so far as the book reveals and fulfills itself completely, it never stops diverting and reversing such a reading, and thus, in a sense, abolishes it" (128).

Less than a decade later, John Johnston (1998) seems to write from a different universe when he analyzes the effects of information explosion on literary texts. Although spatiality is not foregrounded as such in Johnston's analysis, it is everywhere implicit in his notion of "information multiplicity," a vast landscape that, like the cosmos, creates an ever-expanding horizon with no foreseeable limit. A phase change occurs, he suggests, when the separable (and separated) media of *Gravity's Rainbow*, *JR*, and similar texts coalesce into partly connected media systems. Emerging from this dedifferentiation of media come "new behaviors and affective responses that this environment provokes as information becomes completely assimilated in a vast network of media assemblages" (4). Whereas Frank focused on the writer's subjectivity, Johnston, following Deleuze and Kittler, argues that a subject-centered view cannot account for the viral properties of exponentially expanding information: "In the novel of information multiplicity . . . information proliferates in excess of consciousness, and attention shifts to a new space of networks and connections in which uncertainties are structural rather than thematic" (13). The aesthetic unity Frank saw as the principal feature of spatial form now dissolves in the acid bath of information multiplicity: "Negatively defined, a novel can thus be said to become a multiplicity when its fundamental coherence derives neither from a subjective nor an objective unity; that is, when it cannot be adequately defined by the expression of an authorial subject or the totalizing representation of an objective reality" (16).

From unity to assemblage, from subjects who create/apprehend patterns to assemblages that create dispersed subjectivities, from cultural generalizations to technical media as causal agents: these transformations mark the deterritorialized spatial dynamic instantiated and reflected in novels of information multiplicity and media assemblages. "Only a literary form that is machinic, therefore, and which takes the form of an assemblage, can fully register how various information systems, including the mass media, function as part of a larger apparatus of information production and control, while at the same time participating in processes that always exceed them. It is this aspect of information that makes it necessary to consider the novel of information multiplicity as an assemblage produced by a writing machine" (Johnston 1998:14). The writing-down system, in all its technical specificity, thus becomes the source rather than the expression of a conscious subject: "Forms of subjectivity as usually understood are displaced and redistributed through the entire machinic activity that writing and reading entails" (5).

OR simply *assumes* the information explosion that Johnston saw as a formative force on contemporary literature. Information has migrated from a foreground figure where it functioned as a causative agent to the background where it forms part of the work's texture. Whereas Johnston believed that the excess of information could never be contained or halted, *OR* puts information excess into tension with an elaborate set of constraints. It is not excess alone that determines the text's topographic form but rather the interplay between the force information exerts and the constraints that limit and contain it. Moreover, this interplay takes form not merely as a conceptual spatiality (although it has this dimension) but as visual shapes materially present on the pages.

The topographic dimensions are put into play by the configurations of page space and constraints that govern the text's operations. The two narratives center on the "forever sixteen" lovers Sam and Hailey, respectively, with each narrative physically placed upside down on the page and back to front to the other. The turn required to go from one to the other is accompanied by a change in the color coding: Sam is associated with green (green eyes flecked with gold, green ink for every letter *o*), Hailey with gold (gold eyes flecked with green, gold ink for every letter *o*). In the hardcover edition, the color coding is further emphasized by a green ribbon whose top is anchored at the top edge of Sam's narrative, while a gold ribbon is anchored at the top of Hailey's pages. These old-fashioned place markers (which turn out to be remarkably useful) reinforce visually the text's volumetric space, even when the text is closed. The publishers (ventriloquized by Danielewski) recommend that the reader perform the switching operation in units of eight pages of one narrative, then eight pages of the other. This reading practice, which I will call the octet, means that the reader is constantly performing "revolutions" in which the physical book turns 360 degrees (front to back, up to down) each time an octet cycle is completed. Reinforcing the octet are large capital letters at the beginning of each segment, which consecutively form repeating strings of SAMANDHAILEY from Hailey's beginning, and HAILEYANDSAM from Sam's, in anagrammatic fashion.

Narrative, Database, Constraint

At this point, a more fine-grained analysis is needed than Manovich's (2002) rather simple association of narrative with syntagmatic and database with paradigmatic, which, as we saw in chapter 6, is in any event seriously flawed. I identify four different kinds of data arrangements relevant to *OR*, each with its own constraints and aesthetic possibilities.

1. Prohibitions on words and concepts that cannot appear in *OR*. Analogous to paradigmatic substitutions, these prohibitions function as absences that act as constraints and therefore help to define the presences in the text. Particularly important are the clusters printed on the endpapers,

which codetermine what and how verbal representations can appear in the text, as discussed below.

2. Collections of data, which take the form of rotations through a list of possible data elements. Particularly prominent are the lists of plants, animals, and cars that the protagonists encounter (additional substitutions are the minerals associated their lefttwist wristbands). Danielewski wrote an invitation to the users of the *House of Leaves* website to send him the names of "an animal you admire," "a plant you pause for," and "your favorite car," so presumably many, if not most, of these rotational terms come from the data compiled from answers to his invitation.[1]
3. Chronological lists of entries, which form an assemblage in Johnston's sense. Danielewski's invitation also asked for "a specific moment in history, over the last 100 years, which you find personally compelling, defining or at a bare minimum interesting. Necessities: exact data, a refinement of detail, along with a reference or link. An image is also welcome." At least some of the chronological entries can be presumed to come from reader-contributed data. As Danielewski commented in an interview with Kiki Benzon, referring to his invitation, "It's not just my personal history, but histories that go beyond what I can perceive when I'm looking at thousands of books" (Danielewski 2007c).
4. Terms created by permuting a set of elements (for example, letters) through the possible combinations. For example, the name of the manager of the St. Louis café where Sam and Hailey work is variously spelled as Viazozopolis, Viazizopolis, Viaroropolis, etc.

To see how these data arrangements interact with the narrative, consider the topography of the page and page spread. As Danielewski remarks, *OR* activates "a language of juxtaposition" (Danielewski 2007c). Technically free verse, the form is tightly constrained through an elaborate set of topographic patterns (fig. 8.1). Each page is divided into four quadrants. For the left-hand page, the upper left quadrant is the narrative n1 of that page; in the lower left and upside down is the complementary narrative n2 of the other character; in the upper right quadrant is a chronological interval headed by a date (about which more shortly); and in the lower right is the upside-down chronology accompanying n2. For right-hand pages, the same four quadrants apply but with the chronological intervals in the left upper and lower quadrants and n1 and n2 in the right upper and lower quadrants, respectively. The left- and right-hand pages exist in mirror relation to each other, with the chronologies on both pages adjacent to one another across the spine, and the narratives on

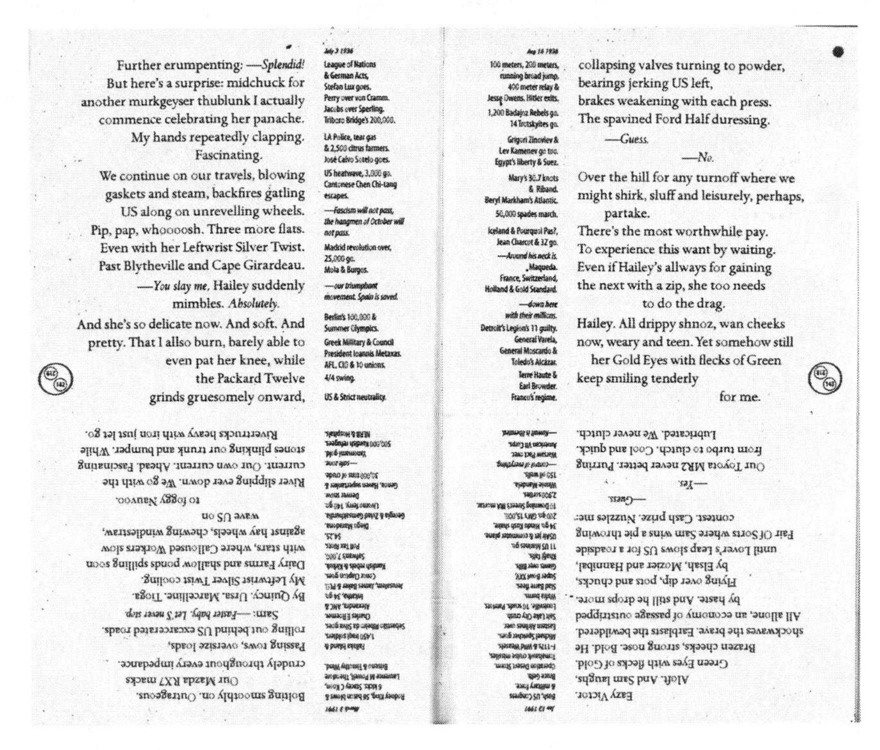

Further erumpenting: —*Splendid!*
But here's a surprise: midchuck for
another murkgeyser thublunk I actually
commence celebrating her panache.
My hands repeatedly clapping.
Fascinating.
We continue on our travels, blowing
gaskets and steam, backfires gatling
US along on unrevelling wheels.
Pip, pap, whoooosh. Three more flats.
Even with her Leftwrist Silver Twist.
Past Blytheville and Cape Girardeau.
—*You slay me*, Hailey suddenly
mimbles. *Absolutely*.
And she's so delicate now. And soft. And
pretty. That I allso burn, barely able to
even pat her knee, while
the Packard Twelve
grinds gruesomely onward,

July 3 1936
League of Nations & German Acts,
Stefan Lux goes.
Perry over von Cramm.
Jacobs over Sperling.
Triboro Bridge's 200,000.
LA Police, tear gas & 2,500 citrus farmers.
José Calvo Sotelo goes.
US heatwave, 3,000 go.
Cantonese Chen Chi-tang escapes.
—*Fascism will not pass, the hangmen of October will not pass.*
Madrid revolution over, 25,000 go.
Mola & Burgos.
—*our triumphant movement. Spain is saved.*
Berlin's 100,000 & Summer Olympics.
Greek Military & Council President Ioannis Metaxas.
AFL, CIO & 10 unions.
4/4 swing.
US & Strict neutrality.

Aug 16 1936
100 meters, 200 meters, running broad jump, 400 meter relay &
Jesse Owens. Hitler exits.
1,200 Badajoz Rebels go.
14 Trotskyites go.
Grigori Zinoviev & Lev Kamenev go too.
Egypt's liberty & Suez.
Mary's 30.7 knots & Riband.
Beryl Markham's Atlantic.
50,000 spades march.
Iceland & Pourquoi Pas?, Jean Charcot & 32 go.
—*Around his neck is*, Maqueda.
France, Switzerland, Holland & Gold Standard.
—*down here with their millions.*
Detroit's Legion's 11 guilty.
General Varela, General Moscardo & Toledo's Alcázar.
Terre Haute & Earl Browder.
Franco's regime.

collapsing valves turning to powder,
bearings jerking US left,
brakes weakening with each press.
The spavined Ford Half duressing.
—*Guess.*
—*No.*
Over the hill for any turnoff where we
might shirk, sluff and leisurely, perhaps,
partake.
There's the most worthwhile pay.
To experience this want by waiting.
Even if Hailey's allways for gaining
the next with a zip, she too needs
to do the drag.
Hailey. All drippy shnoz, wan cheeks
now, weary and teen. Yet somehow still
her Gold Eyes with flecks of Green
keep smiling tenderly
for me.

Bolting smoothly on. Outrageous.
Our Mazda RX7 macks
crudely throughout every impedance.
Passing rows, oversize loads,
rolling out behind US excarcerated roads.
Sam: —*Faster baby. Let'S never stop.*
By Quincy. Ursa. Marcelline. Tioga.
My Leftwrist Silver Twist cooling.
Dairy Farms and shallow ponds spilling soon
with stars, where Calloused Workers slow
against hay wheels, chewing windlestraw,
wave US on
to foggy Nauvoo.
River slipping ever down. We go with the
current. Our own current. Ahead. Fascinating
stones plinking our trunk and bumper. While
Rivertrucks heavy with iron just let go.

March 3 1991
Rodney King, 56 baton blows & 6 kicks. Stacey C Koon, Laurence M Powell, Theodore Briseno & Timothy Wind.
Failaka Island & 1,450 Iraqi soldiers.
Sebastião Ribeiro da Silva goes.
Charles E Roemer.
Alexandra, ANC & Inkatha, 34 go.
Jerusalem, James Baker & PLO.
Conor Clapton goes.
Kurdish rebels & Kirkuk.
Safwan's 7,000.
Poll Tax Riots.
$4.25.
Diego Maradona.
Georgia & Zviad Gamsakhurdia.
Livorno ferry, 140 go.
Denver snow.
Genoa, Haven supertanker & 30,000 tons of crude.
—*safe zone.*
Yanomami gold.
500,000 Kurdish refugees.
NLRB & Hospitals.

Jan 12 1991
Bush, US Congress & military force.
Bruce Gelb.
Operation Desert Storm.
Tomahawk cruise missiles, F-117s & Wild Weasels.
Michael Speicher goes.
Eastern Airlines over.
Salt Lake City crush.
Louisville, 10 scuds. Patriots.
Wafra burns.
Siad Barre flees.
Super Bowl XXV, Giants over Bills.
Khafji falls.
11 US Marines go.
USAir jet & commuter plane.
34 go. Hindu Kush shake.
200 go. GM's 15,000.
10 Downing Street's IRA mortar.
2,900 sorties.
Winnie Mandela.
150 oil wells.
—*control of everything.*
Warsaw Pact over.
American VII Corps.
—*Kuwait is liberated.*

Eazy Victor.
Aloft. And Sam laughs,
Green Eyes with flecks of Gold.
Brazen cheeks, strong nose. Bold. He
shockwaves the brave. Earblasts the bewildered.
All allone, an economy of passage outstripped
by haste. And still he drops more.
Flying over dip, pots and chucks,
by Elsah, Mozier and Hannibal,
until Lover's Leap slows US for a roadside
Fair Of Sorts where Sam wins a pie throwing
contest. Cash prize. Nuzzles me:
—*Guess.*
—*Yes.*
Our Toyota MR2 never better. Purring
from turbo to clutch. Cool and quick.
Lubricated. We never clutch.

Figure 8.1 Sample page from *Only Revolutions*. Image used with permission of Mark Danielewski.

the outer edges. Thus a fourfold mirror symmetry is enacted on each page spread, with a left-right reversal along the spine and an up-down reversal along the horizontal midline. The conceptual significance of these mirror symmetries, which exist in different ways within the narrative diegeses, will become apparent shortly.

In addition to the topographic patterns, numerical and other constraints apply to layout, diction, slang, and conceptualization. On each page, there are 90 words per narrative, making 180 narrative words per page and 360 words across the page spread, enacting another variation of "revolution." In addition, there are 36 lines of narrative per page, counting both the right side up narrative (n1) and the upside down one (n2) (the number of lines in each of the two narratives varies from a minimum of 14 to a maximum of 22, but the total always comes to 36). As one narrative grows, the complementary other shrinks, with each largest at its beginning. There are (naturally) 360 pages in each narrative, with the page numbers written upside down to one another, in green and gold respectively, encapsulated in a circle. These constraints, along with others discussed below, cage the wild yearnings of two sixteen-year-old lovers, who escape school, parents, and authorities but remain imprisoned (as well as articulated) by the writing-down system within which they are encapsulated.

Visually this dynamic is represented by a symbol that appears before the title pages of both narratives, a circle (gold for Hailey, green for Sam) within which sit two vertical lines. The circle I take to refer to the "revolutions" in all their diversity, while the two parallel lines represent Sam and Hailey. This may explain why whenever the letter *l* appears in the text, it is always doubled (as in "allso," "allways," etc.), mimetically reproducing their duality within the revolutions performed by the narratives. The vertical lines may also be taken to refer to the pause symbol on electronic equipment; in this sense, Sam and Hailey exist as "pauses" (sequentially indicated by the chronological intervals) during which the text gives accounts of their actions, as well as the historical events listed under the date heading. Manuel Portela (2011:8) further suggests that the circles encapsulating two parallel lines can be understood as eyes, particularly the reader's eyes that actualize the characters, as discussed below.

John F. Kennedy's assassination on November 22, 1963, the event that the narrator of Don DeLillo's *Libra* says "broke the back of the twentieth century" (1991:181), forms the pivot point of the chronologies. As Mark B. N. Hansen (2011) observes in his discussion of *OR* as a print "interface to time and to history" (179), each narrative explodes outward from this point, Hailey's into the future (with the last chronological interval January 19, 2063), Sam's into the past of the Civil War (the first interval in his chronology is November 22, 1863). Although we can only speculate about why the Kennedy assassination should be singled out in this way, we may note that it forms a communal memory for everyone who lived through it. ("Where were you when Kennedy was shot?" is a question defining a generation, precisely those old enough—and young enough—to preserve this momentous event in their living memories.) As Hansen notes, this communal experience indicates the ways in which readers mobilize the resources offered by *OR* to "negotiate between and across narrative and history, phenomenological and objective time" (2011:186). He points out that the manuscript drafts that have been published indicate that Danielewski started by writing narrative and chronology together in continuous blocks; only gradually did he decide to separate them into two distinct typographical areas. Hansen argues that "what Danielewski effectively accomplishes in de-coupling narrative and historical event is a re-potentialising of history on the collective matrix of his readers' rich embodiment" (187).

Since Sam and Hailey's chronologies do not temporally overlap (except for the fetishistic date November 22, 1963), the only spacetime in which the protagonists logically can meet is in the user's reading practices as she flips the book over and over, an action juxtaposing the protagonists in her imagination. Defying temporal logic, each narrative diegesis has the character meeting his or her complement, falling in love, and reenacting the archetypal tale of Romeo and Juliet. Their *full* story, however, is braided together through the octets. If the text were read linearly through from front to back and then back to front, it would literally be a different text than if read through the octets.

The octet is only the beginning of the text's topographical complexity. In addition to the eightfold reading paths on each page spread are other possibilities that emerge from considering the text as a volumetric space, including alternating between narrators for each page and other reading protocols. Portela (2011:18–19, passim) has diagrammed many of these possibilities, and his analysis reinforces Hansen's emphasis on embodied practices as essential for actualizing (in Hansen's term, "concretising") the text. The multiple reading paths, Portela argues, turn the text "into a machine for revealing the mechanisms that make the production of meaning possible" (2011:8), which we may envision (following the opening discussion) as paradigmatic and syntagmatic variations. The idea is that the multiple reading paths serve as analogies exposing the underlying mechanism of linguistic meaning, the networks of semantic, syntactic, and grammatical differences and similarities that collectively enable signification to occur. Variation creates meaning, and variation is also the mechanism creating the text's complex topography by offering many different reading protocols. Portela continues, "Through this device, [*OR*] displays the intertwined mechanics of writing, language and book, and of the novel itself as a printed and narrative genre" (8).

Because the different reading possibilities create permutations that increase exponentially as the variations multiply, the dimensionality of the text in a technical sense far exceeds the two and three dimensions of pages and volumes. Inductive dimension, the mathematical concept relevant here, can be visualized as taking an object and dragging it in one direction to create a line, with a dimension of one. Now suppose the object is dragged in a new direction, the result being a plane with a dimension of two. Dragging an object is analogous to the actions that the eye performs as it traces lines across pages and through the textual volume. If one reads line by line, page by page, as is putatively the case with a conventional novel, the eye traces a series of planes that, correlated together, create a three-dimensional volume. However, *OR* offers many more possibilities for reading paths, and these unfold as a function of time, including alternating between the two narratives back to front and front to back, as well as reading all the chronomosaics as a group and myriad other possibilities. Now imagine each of these possible reading strategies as creating trajectories that crisscross, fold back on themselves, traverse the text's volumetric space in complicated ways, and so forth. Each of these trajectories may be considered as adding another dimension. In this sense *OR* has an extremely high dimensionality, creating a topological surface of great complexity.

Portela further draws an analogy between the text's high dimensionality (my term, not his) as a material artifact and the digitality of alphabetic languages, the very large number of ways in which twenty-six letters may be combined to create different permutations, as well as the ways in which symbols circulate through a culture in different variations and permutations. "Designed as a textual machine, the text shows the abstractness of signs and culture, i.e., the combinatorial nature of discourse and representation" (Portela 2011:8).

Intrinsic to creating the text's complex topography are the chronological entries or "chronomosaics." Written in epigrammatic style, they merely gesture toward the events they reference. Correlations with the narratives are both extensive and elusive. On January 6, 1971, for example, we read "Berkeley hormones. / Russian long hair. / Coco goes" (91/H/1971).[2] January 6 is the date that a group of researchers from the University of California, Berkeley, announced the first synthetic growth hormone. Hailey's corresponding narrative recounts a threesome between Sam, a woman Hailey calls a "Warm Up Wendy's rear" who comes on to Sam, and a reluctant Hailey. Presumably the connection is hormonal: natural in the narrative, synthetic in the chronology. More opaque are "Russian long hair" and "Coco goes." The latter refers to the death of Coco Chanel (January 10, 1971), four days later than the header. If correlated with the narrative, this interval implies the orgy goes on for four days, illustrating how the chronologies function to give the narratives epic scope. This strategy is even clearer for the entries that begin on October 3, 1929 (96/S/1929). "Tuesday" undoubtedly refers to "Black Tuesday," October 29, 1929 (following "Black Thursday," October 24), when sixteen million shares were sold and the US stock market collapsed completely. The popular aphorism describing the consequences of the collapse, "When

America sneezed, the rest of the world caught cold," is reflected in the narrative: "Then little Hailey sniffs and / desnoots; / —*Ahhh Chooooooooooooo!*" (96/S/1929). The thirteen *o*'s, with their connotation of bad luck, are followed by a scene describing Hailey upchucking, as if a worldwide cataclysm can be compressed into her vomiting, which becomes a synecdoche for the world's attempt to purge itself of the excesses of the 1920s.

A complete exploration of the connections between the narratives and chronomosaics would require researching thousands of factoids, a nearly impossible (and certainly tedious) task. In their multiplicity, they gesture toward a vast ocean of data, even as the text's topography puts severe constraints on the brief entries that, as a group, perform as synecdoches for an inexpressible whole. As a literary strategy, the correlation between the narratives and chronological entries points to a dynamic tension between coordination and contingency, epic inflation and narrative irrelevance. Neither wholly tied to the narratives nor completely untied, history wheels alongside the stories, making them more than personal accounts and less than completely allegorized correspondences. The connections that come into focus, such as the examples above, are patterns that emerge from an ocean of data, much as a Google search imparts a partial ordering on an infosphere too vast to comprehend. Hansen draws the correlative implication that this strategy "foregrounds the crucial role played by selection, a role whose significance increases in direct proportion to the increase in archived information available at a given point in history" (2011:184).

Further emphasizing the vast store of data available on the Internet are the correlations between the diction and slang of the narratives and the chronomosaics; at every point, the characters' colloquial language is appropriate to the period. Arguably, such extensive correlations are feasible only when one has digital databases at one's command. Using Google, I was able quickly to locate many day-by-day chronologies that listed events similar to those Danielewski uses. Sites such as the *Historical Dictionary of American Slang* (http://www.alphadictionary.com/slang/) offer search tools that allow one easily to find slang equivalents for words, with usage dates indicated. In this sense as well, data permeate the text through the vocabulary used by the characters.

In addition to evoking the infosphere and establishing correlations with the narratives, the chronologies paint a canvas as vast as the world, in relation to which the individual desires, fulfillments, and disappointments of the characters are contrasted and compared. "Coco goes," cited above, is one entry among thousands documenting deaths around the world—from accidents, disasters, murders, wars, genocides, diseases, and natural causes. The present-tense "goes," locating the death in the context of a particular day, month, and year, constitutes a "here and now" that becomes truly present only when a reader peruses the page. Concatenated with the time of reading are the temporalities of Sam and Hailey, displaced in time relative to one another yet mysteriously interpenetrating through narrative diegesis and occupying the same page space. Altogether, each page incorporates within its topographic dimension no less than five distinct temporalities (Sam's, Hailey's, their associated chronologies, and the time of reading).

At the same time (so to speak), Sam and Hailey exist solely in the present; past and future do not exist for them. In his interview with the *LAist*, Danielewski remarked, "The characters are moving and are oblivious to history. History is enacted through then. They have no awareness of history. They have no memories" (2007a). "You deal with history on some level when you read the book," he continues. The reader's memories, stimulated by the chronomosaics as well as her growing experience of the narratives as reading progresses, provide the glue that joins narrative and history. The characters' pasts and their anticipated futures (which the choral passages from the plants and animals foreshadow) exist for us but not for them.

And to whom, exactly, does "us" refer? The referential ambiguity of this slider enables a triple pun that linguistically concatenates the characters, American citizens, and globally dispersed readers, respectively. Sam and Hailey denote their special bond by "US," often in ways that portray the two of them standing against the world. In another sense, "US" denotes the United States of America, and in yet another, all the text's readers. Thus "US" refers at once to the exclusivity of two lovers preoccupied with each other while the world whizzes by, the national collective of America, and a transnational community of readers stretching across time and place. As a result, "here and now" becomes a catchphrase to indicate a space and time that is anything but self-evident or self-constituted. Instead, what emerges is a spacetime within whose high dimensionality and complex topography the personal merges with the mythic in the narratives, while in the chronologies, the individual merges with the collective, and the national with the transnational.

What Cannot Be Said

Along with the combinatoric possibilities constituted by the physical and conceptual configuration of page space, an arguably even more important set of constraints is articulated by the endpapers. In mirror (i.e., backward) writing, the colored pages (gold for Hailey's end, green for Sam's) announce "~~The/Now Here Found~~/Concordance." As the strikethrough indicates in its play of absence and presence (recalling Derrida's *sous rature*), this "concordance" is far from a straightforward listing of the text's terms.[3] To explore its function and significance, I refer to *House of Leaves* (2000), Danielewski's sprawling hypertext novel that preceded *OR*. Masquerading as a horror novel but with profound philosophical and psychological depth, *House of Leaves* (despite or perhaps because of its complexity) was a runaway best seller. Danielewski, a relatively young writer virtually unknown prior to the brilliant *House of Leaves*, faced a dilemma similar to that confronting Thomas Pynchon after *Gravity's Rainbow* (1974): what do you do for an encore? In Pynchon's case, a fourteen-year hiatus hinted at the struggle of knowing that whatever he wrote would risk failing to measure up to the extraordinary achievement of *Gravity's Rainbow* (1974). After publishing two smaller works (*The Whalestoe Letters* and *The Fifty Year Sword*), Danielewski tackled another large project. His solution to the Pynchonesque problem was ingenious: he would write the mirror text to *House of Leaves*, inverting its dynamics and flipping its conventions.

Consider the inversions. *House of Leaves* is a large prose hypertext; *OR* is a tightly constrained poem. *House of Leaves* uses footnotes to create multiple reading paths, whereas *OR* uses topographical complexity that works through concatenations rather than links. *House of Leaves* is an obsessively inward work, moving in "centripetal" fashion (Danielewski 2007c) to probe the depths of the house, the psychology of the characters, the family tensions, the cultural contexts, and the convoluted histories associated with the house. *OR*, by contrast, moves outward centrifugally, expressing the wild desires of the sixteen-year-old protagonists in joyrides across the country, free of responsibilities and responsive only to their own couplings and hormonal urges.

Beyond these general mirror symmetries is the elaborate set of constraints articulated by the ovals, ellipses, circles, and other "revolutionary" patterns on the endpapers. Each topographic form articulates an ideational cluster. In my parsing of the clusters, they include kinship ("Brood"); media and mediation technologies ("Write"); grammatical parts of speech and language ("Word"); seeing and looking ("Choose"); grace and condemnation ("Grace"); inwardness; interiority; "in" words such as "inalienable," "inane," etc.; gods and religion ("Devotion"); architectural structures and components; and colors.[4] All of these ideational groups are central to *House of Leaves*, as readers familiar with the text will recognize. In the mirror text

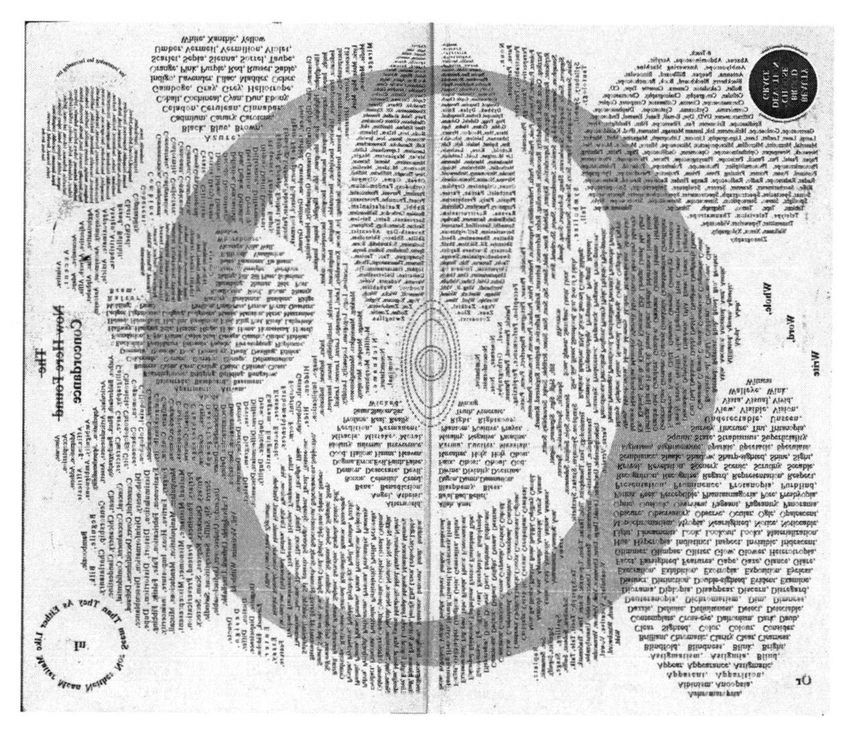

Figure 8.2 Endpapers in mirror writing from *Only Revolutions*. Image used with permission of Mark Danielewski.

of *OR*, they indicate what *cannot be written, cannot be said*. The mirror concordance thus functions as a kind of anticoncordance, indicating words and concepts forbidden in *Only Revolutions*. These metonymic remnants from *House of Leaves*, relegated to the paratextual location of the endpapers and further obscured by appearing as mirror writing, are the paradigmatic equivalents that define the words present in the text by their absences from it (figs. 8.2 and 8.3).

The play between presence and absence intrinsic to paradigmatic variation is a prominent feature of *House of Leaves*. The index to that work, for example, includes entries marked "DNE," which apparently stands for "does not exist." All such entries can, however, in fact be found in the text but in unusual places, such as text in a photograph, words that when elided together form the entry, and other paratextual locations. In *OR*, the paratextual endpapers provide a guide to understanding many of the odd circumlocutions and highly inventive language of the text proper, as important in their textual absence as are the many gaps, holes, and elisions in *House of Leaves*.

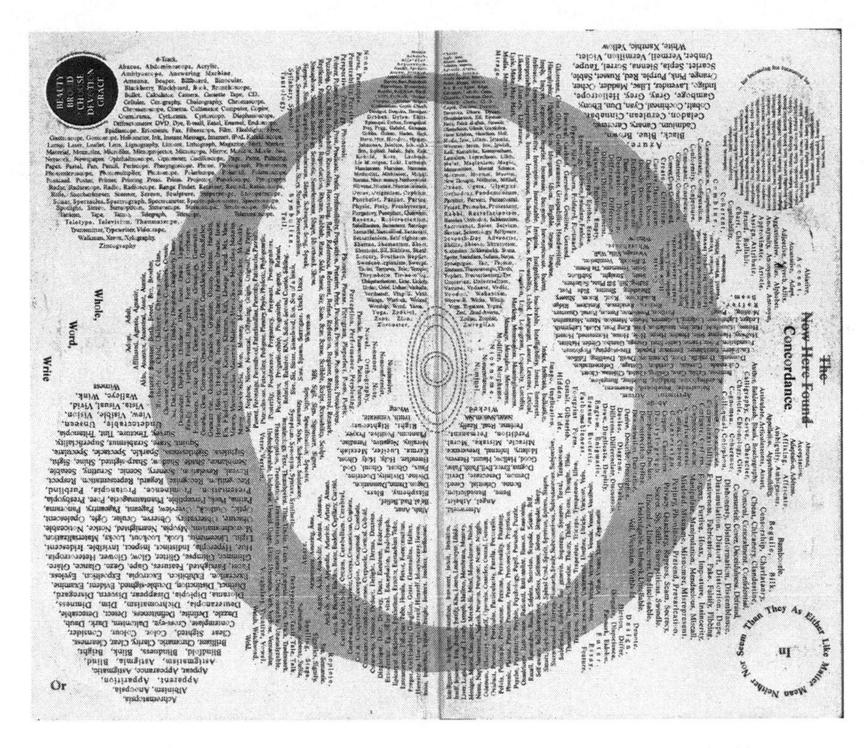

Figure 8.3 Endpapers from *Only Revolutions* flipped in Photoshop so that the type is legible. Image used with permission of Mark Danielewski.

To see how the absences inscribed on the end page clusters help to define the textual presences, consider the relation of some of the clusters to Sam's and Hailey's narratives. Since kinship in all its forms is forbidden, extending even to the prohibition of DNA, Sam and Hailey have no kin—no parents, siblings, or extended family. To all appearances, they are sui generis. Within the narrative, this kinless state correlates with their teenage yearning to be on their own. More constraining from a viewpoint of narrative representation are the prohibitions on interiority. With the exception of Sam's heart, no interior organs are mentioned, nor are there extended probings of their psychological states. When Sam indulges in an orgy, for example, Hailey's distress is shown by the tears she cries, not by explicit reference to her emotions. With psychological rhetoric at a minimum, the inexorable progress of their romance is often articulated through the plants and animals (and sometimes other characters) in Greek chorus–like fashion, issuing warnings, comments, and prophecies for the doomed lovers.

Relevant to the "seeing and looking" cluster is Danielewski's comment that "the word 'light' never appears. . . . Words that are about seeing, for the most part, were taken out. I've been described—not as dogmatic as Oulipo—but there's a resistance to certain things. But the resistance allows for the proliferation of other words" (2007c). In *House of Leaves*, the play between blindness (physical and psychological) and sight/insight is extensive. Zampanò, the putative main narrator, is early on revealed by Johnny Truant to be blind; Will Navidson and others strain to see in the ashen corridors of the house; and an entire chapter is devoted to the biblical twins Jacob and Esau and Jacob's deception of his blind father. Sam and Hailey, despite being on a riotous road trip, never give extended descriptions of the landscape other than allusions to the mountain terrain on which they begin and end their journeys. Seizing center stage are action terms that convey a sense of the landscape not by looking at it but by experiencing it as a three-dimensional topography manifested through movement and velocity. Above all else, the protagonists want to keep moving, expressing their joys (and disappointments) through velocity and speed rather than inward-looking soliloquies.

The media cluster evokes the graphomania of *House of Leaves* and its obsessive interrogation of its own practices of inscription, from the ink spill that obliterates some passages to the Braille encoding, signal-flag symbols, alchemical signs, and myriad other writing practices that fill its pages, including references to film, video, photography, telegraphy, X-rays, radiocarbon dating, and a host of other media technologies. By contrast, technology in *OR* (with the exception of the rotating lists of automobiles in which Sam and Hailey race across the countryside) is almost entirely absent. At the same time, this is an absence that would be almost impossible to achieve without the calculative and data-searching capabilities of networked and programmable machines. As Danielewski acknowledges, "As archaic as [*OR*] is, with its illuminated text and its ribbons, this book could not exist without technology. Without my G5 and 23-inch screen, with two pages on the screen at one time" (2007c).

Moreover, the writing-down system, as Johnston calls it (à la Kittler), includes all of the affordances of the computer, from the Photoshop "reverse image" function that presumably created the mirror writing of the endpapers to the word-count function that was undoubtedly used to create the specified quantities of text on each quadrant, page, and page spread. Because these constraints are central in defining the characters of Sam and Hailey and their expressive possibilities, it is no exaggeration to say, as Johnston anticipates in discussing the novel as media assemblage, "forms of subjectivity as usually understood are displaced and redistributed through the entire machinic activity that writing and reading entails." (1998:5). I would argue, however, that *OR* is finally not a narrative of media assemblage but rather a next-generation form that has gone beyond the shock and awe of first-generation Internet users to bland acceptance of the infosphere as a "natural" part of contemporary life in developed countries. Data flows, unimaginable in their totality, are rendered more or less tractable through increasingly sophisticated search algorithms, mirrored in *OR* through the constraints that partially order and contain information excess. As networked and programmable machines aggregate video, film, sound, and graphics into a single platform, the interplay between text and graphics expands exponentially, as it does in *OR*. In sum, digital-inscription media and the de-differentiation they are presently undergoing can be erased from *OR* precisely because they are omnipresent in its writing practices. The paradigmatic variations, along with mirror symmetries, function as the *visible* linguistic technologies made possible by digital technologies of inscription; nowhere present within the narrative diegesis, digital technologies are everywhere apparent when we consider the writing-down system as a whole.

And what, in this case, is the writing-down system? Once specified by the author, the complex set of constraints become semiautonomous components of it, dictating to the author the spectrum of choices. Cooperating in the authorial project are the software programs, network functionalities, and hardware that provide sophisticated cognitive capabilities, including access to databases and search algorithms. Networked and programmable machines are here much more than a technology the author uses to inscribe preexisting thoughts. They actively participate in the composition process, defining a range of possibilities as well as locating specific terms that appear in the text. The author function is distributed, then, through the writing-down system that includes both human and nonhuman actors.

The distributed author function implies that neither the human creator nor his fictional creatures can credibly claim to be the text's sole author(s). Nowhere within *OR* is the existence of the text itself as a material document explained or inscribed, in sharp contrast to *House of Leaves*, where Johnny Truant tells the story of finding Zampanò's notes and extensively comments on his own writing process, and where the book containing the narratives paradoxically appears within the narrative diegesis. The absence of character-authors in *OR* heightens the importance of the assemblage that forms the writing-down system, visibly apparent on every page, from the historically correct slang, thousands of chronological entries, and elaborate symmetries to the constrained word counts.

The last clusters I will discuss are those centering on gods and religions ("Grace," "Devotion"). Forbidden to articulate "Divine," "Doctrine," "Dogma," "Ghost," "Ghoul," and "God," among other terms, the text shows Hailey and Sam at the beginnings of their narratives as near-demiurges, forces of nature that, while not divine, have exaggerated powers and actions. Sam aggrandizes,

> I'll devastate the World
> No big deal. New mutiny all
> around. With a twist.
> With a smile. A frown.
> Allmighty sixteen and so freeeeee. (1/S/1863)

These exaggerations function as the presences defined by the paradigmatic absences of words more directly evocative of the divine. It is worth noting that within the list of proscribed terms are many antonyms: "Angel" and "Demon," "Paradise" and "Perdition," etc. The inclusion of opposites in many clusters ("Sight" and "Sightless" in the cluster devoted to seeing and looking, for example) indicates that the opposites are engaged in a dynamic of supplementarity, as Derrida would say, mutually defining each other within a cultural context that hierarchically privileges one term as positive, the other as negative. Attending Yale University at the height of deconstruction, Danielewski could scarcely have escaped knowing such academic discourse (Derrida appears in a cameo role in *House of Leaves*, along with other academic stars). The yin/yang-like inclusion of an opposite within the dominant presence of the other term is everywhere apparent, notably in the "Gold Eyes with flecks of Green," and "Green Eyes with flecks of Gold" that appear repeatedly in the text and serve as cover images for the paperback and hardcover dust jacket. The dynamic also works itself out at the level of the narratives, where the hint of death lingers even in the most exuberant expression of life. Indeed, if one is tempted (as I was) to flip the book over when arriving at the ends of the narratives and begin again, this very transformation is enacted as the octets start over in a rhythm that, as the book design hints, is an endless cycle of "only revolutions."

Brian McHale (2011) explores the extent to which *OR* follows Roman Jakobson's insight that literature's "poetic function" is characterized by a transposition of the paradigmatic onto the syntagmatic axis—that is, the overlaying of alterative choices onto the linear order of narrative, a move that deemphasizes the informative and expressive functions and gives priority to "the message as such." Garret Stewart (1990) makes a similar claim about literary language, arguing that its "literariness" comes from a nimbus of homophonic variants activated when a reader subvocalizes the words actually on the page. Many of the literary strategies employed by *OR* create such variants: creative spellings, in which the words inscribed on the page differentially achieve enriched meaning through their relation with the "correct" spellings ("Heart's / pumpin waaaaaay tooooooooo fast" [111/H/1973]); neologisms, evoking the two or more words that they differentially recall; combinatoric variations, already discussed; in a larger sense, the symmetric interplays between Sam and Hailey's narratives, which sometimes perform as paradigmatic variations of one another; and on a meta scale, the mirror symmetries between *House of Leaves* and *OR*. The conjunction of paradigmatic variation with mirror symmetry underscores their similar dynamics, both of which operate as spatial aesthetics. Like paradigmatic variants that haunt the word actually on the page and help to define it through differential relations, mirror symmetry evokes an other at once the same and yet different (in the left-right reversal). Overlaid onto the narrative temporal trajectory, these spatial effects infuse the linear order of syntax with a dense haze of possibilities, as if the words actually on the page operated like electrons. Historically represented as point masses, electrons are now understood to exist as probabilistic clouds that assume specific values only when seen by an observer.[5] Such quantum effects, if I may call them that, are everywhere manifested in *OR*'s linguistic strategies.

Affect and Language

Intimately related to the text's emotional charge is the emergence of the overall temporal patterns. As the two protagonists meet and become lovers, their initial self-centeredness wanes and their immense egos contract to make room for the other, a process expressed visually on the page as the physical space devoted to the narrative shrinks and the other narrative/narrator comes into view as an important force. At the midpoint, each gives to the other equal consideration to the self, signified when each narrative exactly repeats the other, carrying to the extreme the anaphora characteristic of free verse. Significantly, the word at the exact middle of each narrative is "choose," emphasizing the dilemmas that the lovers already sense: leave each other and live, or continue their attachment and die. As the narratives move toward their respective ends, concern for the other supersedes that for the self. Mapping this pattern reveals an "X" structure, in which both protagonists start out at their respective beginnings perceiving themselves as supernaturally empowered and in charge; then, as they open themselves to the other, they begin to experience vulnerability as their growing love for the other gives a hostage to fortune. In the Kiki Benzon interview (Danielewski 2007c), Danielewski remarks, "Freedom is ultimately the quest from anything—to be unrestrained by your circumstances, by your society, by even your own body—whereas love is all about attachment. It's all about the involvement with someone else, which is the opposite of freedom." Yet, as he acknowledges, love (and the bond between Sam and Hailey in particular) has a "transcendent quality. It's through love that you have the greatest amount of freedom." As the two race toward the ending, their foretold fate moves toward tragedy, and, at the same time, the momentum of the octet reading practice catapults them over the ending and into the beginning of a new cycle in yet another "revolution." At the midpoint of this temporal-trajectory-as-circle comes their long hiatus in St. Louis, where they temporarily abandon their road trips as they struggle with the adult responsibilities of earning a living in a hostile environment.

Correlated with this spatialized temporality is the movement of the language. In addition to the narrative slang indexed to the chronomosaics, neologisms, and other linguistic inventions are liveliest and most prolific when the two are on the road, free to express themselves in defiance of decorum and schoolmarm correctness. Checking out a New Orleans band, Sam announces,

> I'm posalutely wild for such
> Cats zesty with slide, ribbing out a
> stride shufflestomping shimsham
> shimmy to time. All mine!
> Toetickling digs, I'm so loose for
> these hands, brillo, di mi, splitticated
> on reed, brass & pluck. Dance. (78/S/1922)

In St. Louis their lives seem to be going nowhere, encaged in alienated labor and subject to the whims of the tyrannical café manager. The language here is correspondingly replete with combinations caged within a tight unyielding frame, for example in the manager's name and the café's title. As if imitating the protagonists spinning their wheels, the language spins through tightly constrained possibilities. These enactments, while not strictly speaking paradigmatic, nevertheless evoke spatialized data arrays (the alphabet envisioned as a string of letters, for example) operating in tension with temporal trajectories.

Complementing the work that the language does in creating hooks for the reader are the symmetries of the plots as they trace the temporal trajectories. One of ways in which the narratives interact with each other, for example, is through ironic contrast. When Sam and Hailey first meet, he announces that she is

> Ashamed she's so slow
> . . . .
> Concerning her poverty,
> I resort to generosity. But
> my offer's too great. She panics.
> Accidentally kicks my nose. (9/S/1870)

Hailey, by contrast, says, "I'm that fast, man," and when Sam tells her, "*Okay, you can be my slave*," she says. "My flying kick nicks his nose. / A warning" (9/H/1963). In other instances, their concatenations are expressed as mirror inversions. Resting in a park, for example, Hailey is approached by a lesbian "GROUNDSLASS" (234/H/1994), while Sam converses with a gay "GROUNDSCHAP" (243/S/1953). Other octets concatenate as similar perspectives on an event, while still others function as complementary halves that together form a whole. When Hailey confesses to Sam she cannot have an orgasm and Sam refuses to ejaculate inside Hailey, for example, their mirror choices indicate psychological reservations about total commitment and therefore limitations on their mutual vulnerabilities. After St. Louis they determine to marry, with or without official sanction. Then, for the first time, Hailey comes and Sam ejaculates inside her, opening them to reproductive possibilities and consequent entry into adult responsibilities. The only way out of this pedestrian future is for them to die, "forever sixteen" and forever free to revel in their unsanctioned pleasures.

As the narratives approach their endings and contract physically on the page, they mimetically reflect not only the deflation of the protagonists' egos (mentioned earlier) but also the narrowing horizons of possibilities for their lives. The motifs that earlier marked the temporal trajectories move toward closure: the twelve jars of honey that they ate along the way and that marked the passage of time have all been consumed; the choruses of plants and animals appear as announcements of species death, although in mirror fashion, since the plants that formerly appeared in Hailey's narrative now populate Sam's and vice versa, their disappearance marked by gray (rather than black) ink; the mountaintops from which Sam and Hailey descended to begin their relationship are inversely reflected in the mountain they scale on their upward climb.

A brief return to aggrandizement has the lovers imagining universal destruction in the wake of their grief for their dead partner. Sam announces,

> How oceans dry. Islands drown.
> And skies of salt crash to the ground.
> I turn the powerful. Defy the weak.
> Only **Grass** grows down abandoned streets, (350/S/1963; boldface in original)

judging that

> No one keeps up and everyone burns and everyone goes.
> I am the big burnout. Beyond speed. (352/S/1963)

As he begins to accept "There is no more way for US. / Here's where we no longer occur" (356/S/1963), the tone modulates as he imagines that some might be responsive to the splendor that was Hailey (no doubt the author's allusion to his hope that readers will experience her death with immersive intensity). Among the many ways in which the "US" that refers to the lovers is concatenated with "US" the nation and "US" the readers is to figure them as outliers who push the boundaries to make sure expansive and expressive possibilities remain for the rest of "US," and it is this tone that dominates at the end: "By you, ever sixteen, this World's preserved. / By you, this World has everything left to lose" (360/S/1963).

With this final turn, the book turns over to begin again, a renewal forecast in the burst of greenery that shoots forth from the icy mountain, foretelling spring, rebirth of young love, and last but not least, the immersive pleasures of narrative amid the topographic dimensions of the text's spatialized aesthetic. *OR* suggests that narrative and its associated temporalities have not gone into decline as a cultural form, as Lev Manovich predicts. Rather, they have hybridized with data and spatiality to create new possibilities for novels in the age of information. As the book turns, and turns again, the title of this extraordinary work broadens its connotations to encompass the dynamic of renewal that, even as it obliterates traditional novelistic form, institutes a new typographical ordering based in digital technologies—revolutionary indeed.

Coda: Machine Reading *Only Revolutions*

By N. Katherine Hayles and Allen Beye Riddell

As a densely patterned work, *OR* lends itself well to machine reading. The extent of the symmetries between Sam and Hailey, the constraints governing word choice, the progression of the narrative, and the correlation of narrative, bibliographic, and semantic codes are not only verified but brought into sharper focus by machine reading. For the most part, our discoveries reinforced what we had already discovered through close reading, but in a few instances, they revealed new information that extended and deepened our understanding of the text.

Our first step was to hand code the entire text[6] and import it into a database, with special categories for cars, plants, animals, minerals, and place-names, and with every word indicated as originating with Sam or Hailey's narratives, respectively. With the place-names identified, we then overlaid them onto a Goggle map. At this point, considerable hand-correction was necessary, since many of the place-names ("Rochester," for example) had possible identifications in several different states. The results are shown in figures 8.4, 8.5, and 8.6. The surprise here is the extension of their westward journey to Hawaii and Alaska. These place-names occur in the context of a global traffic jam that they cause, as referenced in Hailey's narrative (Sam's has a parallel passage, where it is conceived as a wedding present to Hailey):

> Screeching to a standstill. Barring behind US all
> transport modes. Junkers, Grimmers and Tam Bents
> turned here to an impound lot. Dead
> Mulberries & Morels ignored by every horn.
> And no wheel can pass these wheels.
> Sam's Jeep Gluon flung across every start.
> From Bangor[7] to Los Angeles by
> Barrow to Wailuku.
> A globally hubbed hork.
> (299/H/8 March 2015–300/H/15 November 2015)

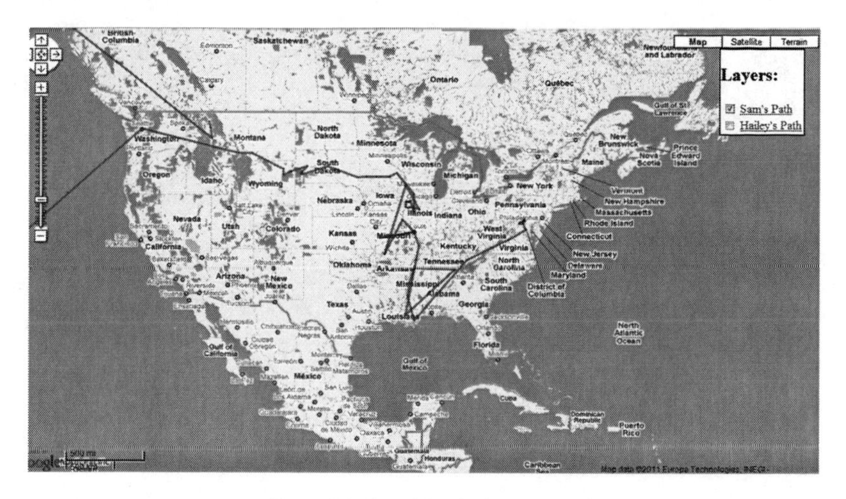

Figure 8.4 Map of Sam's place-names.

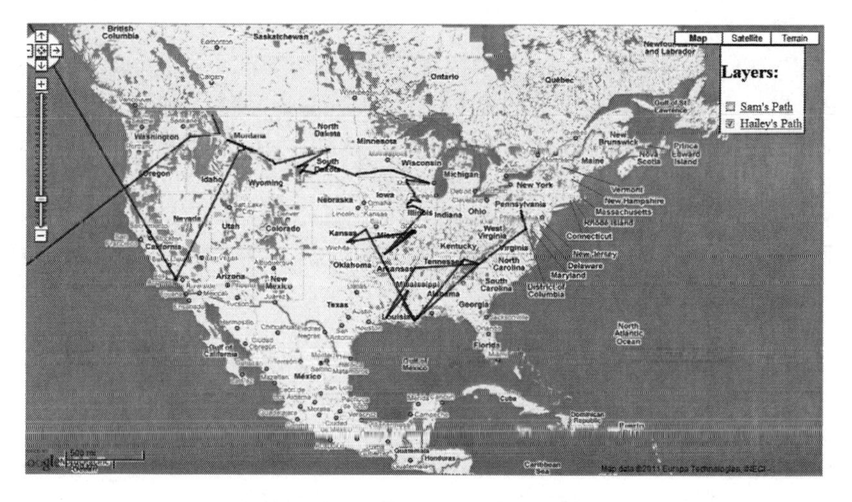

Figure 8.5 Map of Hailey's place-names.

The effect in the narrative is to suspend all traffic as it piles up in a "Transapocalyptics" (301/H/15 November 2015). (The dates indicate that the jam lasts for eight months!) In the *LAist* interview, Danielewski remarks of Sam and Hailey, "What's terrifying about them is that the world withers and shakes and burns to the ground around them, but it doesn't bother them at all. They are so caught up in their affection for each other and their antics that they lose track" (2007a). The global chaos implies that the rest of the

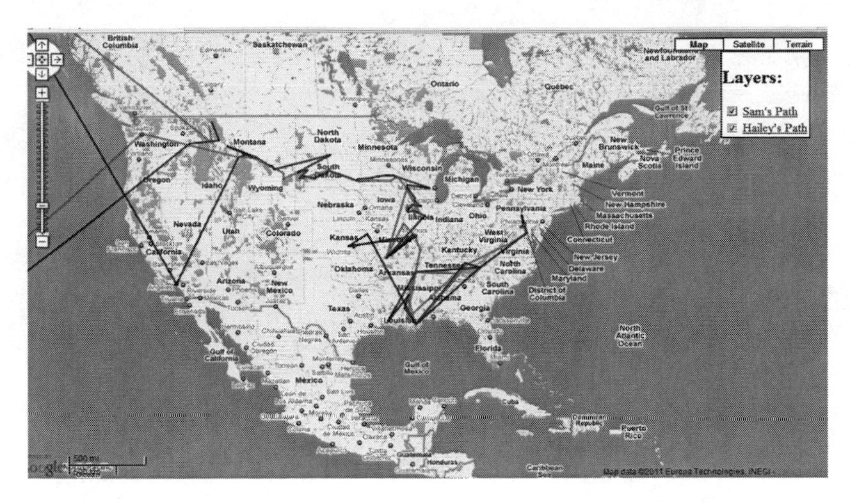

Figure 8.6 Sam's and Hailey's place-names juxtaposed.

world is stopped, jammed to a halt, while only the lovers are free to race ahead and keep moving. When their vehicle spins off the road on ice, however, they are thrown clear and continue on foot up the mountain, where they will meet their respective ends as their wild road trip comes to an end. Nevertheless, the logic of the octet, in which the end of one narrative is the beginning of the other, leaps across this stasis and converts it into a caesura, a temporary pause before the action starts all over again.

Another aspect of the geography of Sam and Hailey's cross-country spree is the numerical coding of its georeferenced points. Portela (2011:53) points out that the dip south into New Orleans follows the longitude 90 degrees west and that the journey west to the mountain largely occurs on Interstate 90. As a factor of 360, 90 connotes a quarter circle; since the 90-degree turn south is followed by a 180-degree turn north and another 90-degree turn west, the complete trajectory forms a 360-degree circle, thus making good the road trip as another manifestation of a "revolution."

Further insight is gained by comparing the word frequency of Hailey's narrative with that of Sam's. The extensive parallels between the two narratives are borne out by the correspondence between paired frequency counts. For example, "now," a word freighted with significance given the temporal separation between the characters, occurs 111 times in both narratives, marking a perfect symmetry between the two and conveying a shared emphasis on living in the present. "Allways," another highly significant term, occurs 111 times in Hailey's, 110 times in Sam's. "Allways" echoes the extensive exploration of "hallway" in *House of Leaves*, making good another mirror correspondence; it also hints at a projection into the future that will not occur in one sense (since the protagonists die) but that is accurate in another sense (because the front-back symmetries enjamb the ending together with the beginning, and more generally because literary narratives are typically taken in critical analysis to exist in an omnipresent present tense). "Never" occurs 72 times in Hailey's and 70 times in Sam's, marking a symmetry that serves to reinforce their mutual vows to one another. Even such nonsignificant words as "and," "the," and "of" show a close correspondence between the two narratives. As noted earlier, Danielewski composed with the two narratives side by side, with corresponding pages both displayed on the monitor, and the parallelism is carefully structured to show the growing love between the protagonists, who initially start far apart, come closer until their residence in St. Louis (the midpoint of the narrative), and then follow each other in declining action until their respective deaths.

Perhaps the most illuminating discovery comes from a hint contained in the "spoiler" poster published in *Revue Inculte* 14 (Danielewski 2007d), a large colored print object measuring about two feet by three feet, constructed from notes that Danielewski prepared for his French translator (see our digitized version with zoom function at http://onlyrevolutions.info/). Along with tables showing the extensive numerical parallels between Sam and Hailey and the factors of 360 that govern the division of the narratives into sections, pages, paragraphs, and lines are schematics of the chronological intervals associated with each narrative, showing again in the sidebars the carefully planned symmetries between the two chronologies. Across the poster's lower half are the thematic clusters worked out through the different sections, indicating the meticulous and extensive planning that went into the narrative progressions. Also included are diagrams showing the narrative progression of each character's "disregard for" (declining) and "esteem for" (rising) the other as the narrative proceeds. A chart of "differences of perception" shows that they start far apart, meet at the narrative midpoint, then diverge again as each character places the other above care for himself or herself.

Informative as the poster is about narrative structures, word choices, and chronologies, it is reticent on one point: included is a large column labeled "Nix List" that has been blanked out, suggesting that Danielewski provided his French translator not just with conceptual clusters but specific words that he wanted not to appear in the translated text. This presents an intriguing problem: how do you find the words that are not there? Our solution is to compare the word frequencies in *OR* with the Brown corpus, a database of one million words carefully selected to be statistically representative of twentieth-century American prose. To assist our comparison, we calculated a chi-square statistic for each word in *OR*, which provides a rough measure of how noticeable the difference is between the observed frequencies in *OR* and the Brown corpus.[8] For example, *in*, one of the semantic values of the "inwardness" cluster, appears 21,402 times in the Brown corpus, 0 in *OR*; *into* occurs 1,789 in Brown and 0 in *OR*. Other nonoccurring words with high chi-square values are *was*, *were*, and *been*; it is remarkable that these past tenses of *to be*, the most frequent verb in English prose, are entirely absent in *OR*, a choice (and a constraint) emphasizing that the protagonists live vibrantly in the present. Also absent are *as* and *like*, indicating a preference for neologisms and adjectives over similes or analogies, a choice that vivifies the action and refuses the reflective pauses (for the narrator and reader) associated with such literary tropes as epic similes. *They* and *people* are also absent, showing the lovers' disregard for the social collectives that would restrain their freedom and delimit their choices. *Said* also does not occur, showing the lack of a narrator who reports on the lovers, as distinct from them speaking their own thoughts and the Greek-chorus comments of the animals and plants.

One puzzling nonoccurring word is *or*, represented on the endpapers as its own circle enclosed with a series of ellipses consisting of very tiny repetitions of "or." We conjecture that *or* is forbidden because it is the acronym that Danielewski (and others) typically use for *OR*. One of the forbidden clusters has do to with self-reflexivity, in the sense that the text cannot refer to the textual components prominent in its composition, such as *novel*, *pattern*, and *poem*. Thus *self-reflexivity has been banished from the semantic register and displaced onto the topographic*, another indication of how important the spatial aesthetic is to this text. As discussed earlier, the text's topographic complexity serves as an analogy that extensively interrogates the text's existence as a novel, a linguistic artifact, and a material object instantiated in a codex. In the *LAist* interview, Danielewski commented, "A lot of *Only Revolutions* is interested in the mechanisms that are underlying things . . . the grammar, the physics of things. We're not talking about particular words but the relationship between words. Not the particular names of planets, but the nature of an ellipse and the effect of gravity on the orbit" (2007a).

We believe it is fitting that we use digital computers to analyze *OR*. In a certain sense, as Portela observes, *OR* employs the resources of the codex as an aesthetic, bibliographic, and material form to exploit the digital nature of alphabetic language. In this sense, the numerical codes implicit in numbers of words, lines, paragraphs, pages, and sections compute the text's own conditions of possibility. Forbidden to refer to itself through the semantic register of *or*, *OR* nevertheless functions as a series of recursive algorithms whose operations produce the text as a print artifact and as a two- and three-dimensional object, while simultaneously inviting readers to increase its dimensionality exponentially through the multiple reading paths and page symmetries it offers. Commenting on the text's deep recursivity and numerical codes, Portela argues that "*Only Revolutions* links the digitality inherent in human language and in alphabetic writing, as permutational devices based on recursive structures, to the system of differences that sustain the material and conceptual space of the codex. . . . Instead of the common figure of the computer as a book, i.e., as an extension of the informational structure of the codex, Danielewski's work gives us the book as a computer, i.e., as a calculating machine that generates algorithms and geometrizes the plane and the space of writing and reading" (2011:71) With this coda, *How We Think: Digital Media and Contemporary Technogenesis* concludes with an instance of technogenesis redefining the codex as a digital technology that, in cycles of continuous reciprocal causation, both influences and is influenced by the functionalities of networked and programmable machines. To grasp fully the dynamic now in play between print forms and digital technologies, we must consider them as mutually participating in the same media ecology.

Time and space, narrative and database, institutional programs and the history of digital and print technologies are the sites that have been explored in this book that, in recursive fashion, partakes of digital media even as it also reflects the practices of print scholarship. The rich conceptualizations and intricate patterns of *TOC*, *RST*, and *OR* show that technogenesis has a strong aesthetic dimension as well as neurocognitive and technical implications. They demonstrate that in this cultural moment fraught with anxieties about the future, fears for the state of the humanities, and prognostications about the "dumbest generation," remarkable literary works emerge that can catalyze audiences across the generations. These works vividly show that the humanities, as well as our society generally, are experiencing a renewed sense of the richness of print traditions even as they also begin to exploit the possibilities of the digital regime. In my view the humanities, far from being in crisis, have never seemed so vital.

Notes

Chapter One

1. I do not mean to imply that books are obsolete. On the contrary, book art, culture, and experimentation are experiencing renewed bursts of interest and innovation. What I do mean to imply is that print is no longer the default medium of communication. I believe that there is a strong causal relationship between print's coming into visibility as a medium and the robust interest in it as an aesthetic and artistic form.

2. For a broad spectrum of views on the question "Is the Internet changing the way you think?" see Brockman (2011).

3. Jessica Pressman and I are in the process of creating a reader demonstrating the range and potential of Comparative Media Studies, with essays by distinguished scholars across the full range of media forms, from scrolls to computer games. We anticipate publication in late 2012.

4. *Umwelt* of course refers to the biosemiotics of Jakob von Uexküll, most famously represented in the example of a tick. If a tick—blind, limited in motion, and having only one goal in life (to suck mammal blood)—can have an umwelt, then surely an object such as Rodney Brooks' "Cog" head-and-torso robot (http://www.ai.mit.edu/projects/humanoid-robotics-group/cog/) can have one. Cog has saccading eye tracking, hand and head movements, and a variety of other sensors and actuators. Compared to a tick, it's a genius.

Chapter Two

1. Stephen Ramsay (2008b) follows the mainstream view in identifying its origins with the work of the Jesuit priest Robert Busa in the 1940s; Willard McCarty (2009) suggests "that machine translation . . . and information theory . . . are the beginnings of humanities computing."

2. Willard McCarty (2009) prefers "humanities computing" precisely because it emphasizes the computational nature of the field as he practices it; he recognizes that there is a significant difference between text analysis and projects such as, for example, Bernard Frischer's *Rome Reborn*.

3. For the possibilities and challenges of partnering with museums and other art institutions, see Schnapp (2008).

4. Willard McCarty (2009), stressing the interaction between machine analysis and human reading, suggests that the result is "reading in which the machine constitutes a new kind of book [that] we don't even have the vocabulary to describe."

5. Models, simulations, and correlations overlap but are not identical. As Willard McCarty (2008) points out in his analysis of modeling, models come in at least two main types: physical objects that capture some aspects of systems, and mathematical representations that (sometimes) have predictive power about how systems will react under given sets of circumstances. Simulations have similarly fuzzy boundaries, ranging from a set of computer algorithms that operate within a defined environment to play out various possibilities to simulations that combine real-time input with programmed scenarios (the simulations the Institute for Creative Technology has created for army training purposes are examples). In complex dynamical systems, simulations frequently are the tools of choice because the recursive interactions and feedback loops are too complex to be modeled through explicit equations (for example, simulations of the weather). Correlations of the kind that Franco Moretti (2007) uncovers are perhaps the least precise, in the sense that the causal mechanisms are not explicitly revealed or implied and must be arrived at by other means.

6. Philip Ethington, e-mail message to author, April 14, 2009.

7. Response to a presentation of "How We Think," Los Angeles, CA, June 1, 2009.

8. See Presner (2009a) for a discussion of the military technologies on which Google relies and the erasures Google Maps and Google Earth impose.

9. There are, of course, many scholars working in these areas, notably Wendy Hui Kyong Chun (2008), Lisa Nakamura (2007), and Thomas Foster (2005).

Chapter Three

1. Addie Johnson and Robert W. Proctor, in *Attention: Theory and Practice* (2004) track the growth of articles in psychology on attention (3). The number of articles with *attention* in the title published in jumped from 333 in 1970 to 1,573 in 1990. By the year 2000, the number of articles published had swelled to 3,391. Tracking *attention* as a keyword shows a similar increase, from 8,050 instances in 1990 to 17,835 in 2000.

2. Researchers in the field of attention studies identify three major types of attention: controlled attention, capable of being focused through conscious effort; stimulus-driven attention, a mode of attentiveness involuntarily attracted by environmental events, such as a loud noise; and arousal, a general level of alertness (see Klingberg [2009:21] for a summary). In these terms, deep attention is a subset of controlled attention, and hyper attention bridges controlled and stimulus-driven attention.

3. For a contrary view, see Bernard Stiegler, *Taking Care of Youth and the Generations* (2010). Citing my work on hyper and deep attention, Stiegler argues that the phenomenon

is much graver than I acknowledge; rather, attention is actively being *destroyed* by what he calls the audiovisual (i.e., film and television) and programming industries. The following passage is typical: "This partnership [between the audiovisual and programming industries] had precipitated a set of conflicting forces, *attentional deficiencies* brought about by psychotechnical attention capture, whose current result is an immense psychological, affective, cultural, economic, and social disaster" (2010:58; emphasis in original). He criticizes my work for its emphasis on pedagogy as inadequate to capture the gravity of the crisis, arguing that nothing less than Western democracy and rational thought are at stake. He also argues in relation to my term "hyper attention" for a distinction between attention as duration (attending to an object for a long period of time) and as concentration (the intensity of attention). In his terminology, what is important are "the lengths of the circuits of transindividuation [attention] activates. . . . Each circuit (and its length) consists of many connections that also form a network, as another constituent of depth, a kind of texture" (2010:80). Although I do not agree with his broad condemnations and generalized conclusions, I find his study useful for its rethinking of the philosophic tradition in terms of attention and for his analysis of the connections between attentive focus and transindividuation.

Chapter Four

1. I am speaking here of literal organs, not the "body without organs" made famous by Gilles Deleuze and Félix Guattari in *A Thousand Plateaus: Capitalism and Schizophrenia* (1987).

Chapter Five

1. In researching the archival material discussed this chapter, I relied primarily on four repositories of telegraph code books: (1) The Fred Brandes Telegraph Code Book Collection; references to these books are indicated by FB. (2) The telegraph code collection at the National Cryptologic Museum Library; works from this source are referenced by NCM. (3) The website of John McVey, referenced as JM (http://www.jmcvey.net/cable/scans.htm). (4) The Nicholas Gessler "Things That Think" collection; material accessed here is referenced by NG.

2. Phrases including "economy," simplicity," and "secrecy" are ubiquitous; see for example *The ABC Universal Commercial Electric Telegraph Code* (1881:1).

3. This phrase appears over and over in comments about the telegraph, especially the electric telegraph. It also appears frequently in telegraph code books; see for example *The ABC Universal Commercial Electric Telegraphic Code* (1881): "Every day the value of the telegraphic system is being more and more felt. It has revolutionized our commerce, and proved of inestimable value in domestic affairs; there is no relation of life in which the influence of the change it has effected has not been felt. Electricity had done as much as human effort can accomplish towards the annihilation of space and time, and its adaptation for practical purposes is, perhaps, the greatest invention of the present century" (1).

4. This anecdote was related to Nicholas Gessler at the 2009 Bletchley Park Conference on Alan Turing, which was attended by some of the women who had worked at Bletchley during World War II.

5. Frank Darabont's contemporary novel *Walpuski's Typewriter* (2005) takes this idea to the extreme, conjuring a demon-infested machine to which one need only feed paper to have it spew out best selling potboilers.

6. The method is similar to the alternation of numbers and letters in Canadian zip codes, which once saved me from mistaking a hastily scribbled 5 for an *S*.

7. John Durham Peters (2006) offers a useful correction to Carey, pointing out that the separation of transportation and communication had already taken place to some extent with semaphoric and optical telegraphs before the advent of the Morse electric telegraph. Nevertheless, it is fair to say that the Morse telegraph and its competitors in electric telegraphs made this phenomenon much more widespread and common than it was before.

8. In fact, most of the versions of "O Susanna" available on the web omit this verse altogether and add others that Foster did not in fact write. The original is available at http://www.archive.org/stream/stephencollinsfo00millrich/stephencollinsfo00millrich_djvu.txt.

9. I am grateful to Nicholas Gessler for assistance in researching this background.

10. Such a view has, of course, been effectively critiqued by many cultural critics, including, in addition to Raley, Gayatri Spivak and Lydia Liu, among others. Since English is typically the dominant language into which other languages are translated, the idea of code as universal language is often associated with the forging of a special relationship between code and English, for example by the spread of Global English through English-based programming languages. Many international software companies, when working in countries with nonalphabetic languages such as China, Japan, and South Korea, have found it easier to teach English to their employees (or more commonly, require them to know English as a condition of employment) than to use ideogrammatic characters in programming. Positioning code as a universal language not only erases cultural specificities but also creates a situation in which a dominant language (English) tends to suppress the cultural views and assumptions embedded in nonalphabetic or "minor" languages.

11. See for example the elegant Bion silver cipher disk in the Gessler collection, http://www.duke.edu/web/isis/gessler/collections/crypto-disk-strip-ciphers.htm.

Chapter Six

1. The exception is the null value, which has its own problems as discussed above.

2. Jerome Bruner (1992) also emphasizes the importance of causality in discussing narrative, identifying crucial components as agency, sequential order (or causality), sensitivity to the canonical (or context), and narrative perspective (1992:77).

3. The Spatial History Project describes the process of overlaying USGS quads onto a Google Earth image as "georeferencing." The more typical sense of the term refers to locating an object in physical space, for example locating objects in an aerial photograph on a map. The Google Earth image onto which the spatial history project overlays USGS quads is itself a satellite representation, so placing it onto a map would be the usual way in which georeferencing takes place, rather than georeferencing the USGS quads to a Google image. It is worth noting that the USGS quads mapped after about 1920 used not only on-the-ground surveys but also aerial photographs, so in a sense, they already contain the information that would be created by georeferencing them to a Google Earth image.

4. ArcGIS is a proprietary software system marketed by ESRI. Although ArcGIS is not the only GIS system available, it is by far the most widely used, analogous in this sense to the Microsoft operating systems. Like Microsoft, ArcGIS is very much a corporate product, aggressively marketed through seminars, competitions, etc.

5. I am indebted to Allen Beye Riddell for drawing my attention to this programming technique.

Chapter Seven

1. As Jessica Pressman (2009) notes, "Mycroft Ward" is a near-homophone for "Microsoft Word"; like the ubiquitous software, Ward threatens to take over the planet. Mycroft is also

the name of Sherlock Holmes's elder brother, described in Arthur Conan Doyle's "The Adventure of The Bruce-Partington Plans" ([1908] 1993) as serving the British government as a kind of human computer: "The conclusions of every department are passed to him, and he is the central exchange, the clearinghouse, which makes out the balance. All other men are specialists, but his specialism is omniscience" (766). This, combined with his distaste for putting in physical effort to verify his solutions, makes him a suitable namesake. The reference is reinforced when Clio tells the First Eric that when she was hospitalized for cancer, she read Doyle stories until she was sick of them.

2. The name Ward, and the reference to him as a "thing," suggest a conflation of two H. P. Lovecraft stories of body snatching and mind transfer, *The Case of Charles Dexter Ward* ([1941] 2010) and "The Thing on the Doorstep" ([1937] 1999). Certainly, there is a Lovecraftian dimension to the fusion of technoscience and the occult in Ward's scheme. I am indebted to Rob Latham for drawing this parallel to my attention.

3. We are told that Mycroft Ward wants to take over the Second Eric because it knows that the Ludovician is hunting him and it needs the Ludovician in order to expand its standardizing procedure without limit (S. Hall [2007] 2008a:282–83). Although we are not told why the Ludovician would enable this expansion, it seems reasonable to conclude that the personalities of the node bodies put up some resistance to being taken over, and overcoming this resistance acts as a constraint limiting the number of node bodies to about a thousand. Apparently the idea is that the Ludovician will be used to evacuate the subjectivities that Ward wants to appropriate, annihilating the resistance and enabling Ward's expansion into the millions or billions.

4. Hall follows the convention of omitting periods after Mr, Dr, etc.

5. See Barbara Hui (2010:125) for a mapping of these locations.

Chapter Eight

1. As a registered member of the website, I received this e-mail, from which I am quoting.

2. The notation indicates the page number, the narrator (Sam or Hailey) and, since the language is synchronized with the vocabulary current on the indicated date, the chronological heading to provide historical context.

3. The words "The/Now Here Found" are struck because they appear in the text; they are thus in a different category than the rest of the words on the endpapers. "Concordance" is not struck through because it does not appear in the text proper; neither do all the words on the endpapers other than those few in special locations, as noted in the following discussion.

4. The words "Beauty," "Brood," Choose," "Devotion," and "Grace" appear in the circle with a black background, white perimeter, and black lines striking through tiny red words, "Found Once, Once Here" on one side, and on the other, "Found Once, Once There." As this cryptic message suggests, these words, in addition to naming some of the categories, are found once and only once in each narrative of Sam and Hailey (hence "Found . . . Here" and "Found . . . There").

5. For a summary of quantum effects as they are currently understood, see Tom Siegfried, "Clash of the Quantum Titans" (2010).

6. Our thanks to Abraham Geil for performing the laborious task of entering every word into a spreadsheet.

7. In the interest of clarity, we did not map Bangor, Maine, in our representation.

8. The chi-square test (from which the statistic gets its name) is proposed as a test to determine whether two samples come from the same theoretical distribution. While the test requires a number of assumptions that are unmet here, the statistic itself offers a serviceable indicator of how divergent are two observed word frequencies. For a fuller discussion of methods of comparing two corpora, see Kilgarriff (2001).

Works Cited

Abbreviations

FB Fred Brandes Telegraph Code Book Collection
JM John McVey website, http://www.jmcvey.net/cable/scans.htm
NG Nicholas Gessler "Things That Think" collection
NCM National Cryptologic Museum Library telegraph code collection

The ABC Universal Commercial Electric Telegraphic Code. 1901. 5th ed. FB. London: Eden Fisher.

Ambrose, Stanley. 2001. "Paleolithic Technology and Human Evolution." *Science* 291.5509 (March 2): 1748–53.

American Radio Service Code No. 1. Signed by James W. McAndrew, chief of staff under General Pershing. NG.

Amerika, Mark. N.d. Alt-X Online Network. http://altx.com/home.html.

Amin, Ash, and Nigel Thrift. 2002. *Cities: Reimagining the Urban*. London: Polity.

Amiran, Eyal. 2009. Interview with N. Katherine Hayles. Los Angeles, CA, and Hillsborough, NC, January 8.

Artillery Code, 1939: Instructions for Encoding and Decoding. 1939. NG.

N.d. *"BAB" Trench Code No. 4*. FB.

Bal, Mieke. 1998. *Narratology: Introduction to the Theory of Narrative*. 2nd ed. Toronto: University of Toronto Press.

Baldwin, James Mark. 1896. "A New Factor in Evolution." *American Naturalist* 30.354 (June): 441–51.

Balsamo, Anne. 2000. "Engineering Cultural Studies: The Postdisciplinary Adventures of Mindplayers, Fools, and Others." In *Doing Science + Culture: Why More Americans Are Doing Wrong to Get Ahead*, edited by Roddey Reid and Sharon Traweek. New York: Routledge. 259–74.

———. 2011. *Designing Culture: The Technological Imagination at Work*. Durham, NC: Duke University Press.

The Baltimore and Ohio Railway Company Telegraphic Cipher Code. 1927. May 14.

Bamboo Digital Humanities Initiative. http://projectbamboo.org/.

Bargh, John A. 2005. "Bypassing the Will: Toward Demystifying the Nonconscious Control of Social Behavior." In *The New Unconscious*, edited by Ran R. Hassin, James S. Uleman, and John A. Bargh. Oxford: Oxford University Press. 37–60.

Bates, David Home. 1907. *Lincoln in the Telegraph Office: Recollections of the United States Military Telegraph Corps during the Civil War*. New York: Century.

Bauerlein, Mark. 2009. *The Dumbest Generation: How the Digital Age Stupefies Young Americans and Jeopardizes Our Future*. New York: Jeremy P. Tarcher/Penguin.

Bear, Mark F., Barry W. Connors, and Michael A. Paradiso. 2007. *Neuroscience: Exploring the Brain*. Baltimore: Lippincott Williams and Wilkins.

Benjamin, Walter. 1968a. "The Task of the Translator." In *Illuminations*, edited by Hannah Arendt, translated by Harry Zohn. New York: Schocken Books. 70–82.

———. 1968b. "The Work of Art in the Age of Mechanical Reproduction." In *Illuminations*, edited by Hannah Arendt, translated by Harry Zohn. New York: Schocken Books. 217–51.

Bentley, E. L. (1929) 1945. *Bentley's Second Phrase Code*. New York: Rose.

———. 1902. *Bentley's Complete Phrase Code*. New York: Rose.

———. 1945. *Bentley's Complete Phrase Code*. New York: Rose.

Bergson, Henri. (1913) 2005. *Time and Free Will: An Essay on the Immediate Data of Consciousness*. Chestnut Hill, MA: Adamant Media.

Best, Stephen, and Sharon Marcus. 2009. "Surface Reading: An Introduction." *Representations* 108.1 (Fall): 1–21.

Blackmore, Susan. 2010. "Dangerous Memes; or, What the Pandorans Let Loose." In *Cosmos and Culture: Cultural Evolution in a Cosmic Context*, edited by Steven Dick and Mark Lupisella. Washington, DC: NASA History Division. 297–318.

Bobley, Brett, Jason Rhody, and Jennifer Serventi. 2011. Interview with N. Katherine Hayles. Washington, DC, May 9.

Bodenhamer, David. J., John Corrigan, and Trevor M. Harris, eds. 2010. *The Spatial Humanities: GIS and the Future of Humanities Scholarship*. Bloomington: Indiana University Press.

Bogost, Ian. 2007. *Persuasive Games: The Expressive Power of Video Games*. Cambridge, MA: MIT Press.

———. 2009. Response to "How We Think: The Transforming Power of Digital Technologies." Presentation at Georgia Institute of Technology, Atlanta, GA, January 15.

Bol, Peter. 2008. "Creating a GIS for the History of China." In *Placing History: How Maps, Spatial Data, and GIS Are Changing Historical Scholarship*, edited by Anne Kelly Knowles. Redlands, CA: ESRI Press. 27–59.

Bolter, Jay David. 2008. Interview with N. Katherine Hayles. Atlanta, GA, and Hillsborough, NC, October 13.

Bowker, Geoffrey C., and Susan Leigh Star. 2000. *Sorting Things Out: Classification and Its Consequences*. Cambridge, MA: MIT Press.

Brockman, John, ed. 2011. *Is the Internet Changing the Way You Think? The Net's Impact on Our Minds and Future*. New York: Harper.

Brooks, Fred. 2010a. *The Design of Design: Essays of a Computer Scientist*. Reading, MA: Addison-Wesley Professional.

———. 2010b. "The Master Planner: Fred Brooks Shows How to Design Anything." Interview by Kevin Kelly. *Wired* 18.8: 90, 92.

Bruner, Jerome. 1992. *Acts of Meaning: Four Lectures on Mind and Culture*. Cambridge, MA: Harvard University Press.

Burns, Arthur. 2009. Interview with N. Katherine Hayles. Centre for Computing in the Humanities, King's College, London, January 23.

Byrn, Edward W. 1900. *The Progress of Invention in the Nineteenth Century*. London: Munn and Munn.

Carey, James W. 1989. "Technology and Ideology: The Case of the Telegraph." In *Communication as Culture: Essays on Media and Society*. New York: Routledge. 201–30.

Carlson, Andrew, et al. 2010. "Toward an Architecture for Never-Ending Language Learning." *Proceedings of the Twenty-Fourth AAAI Conference on Artificial Intelligence*. http://www.aaai.org/ocs/index.php/AAAI/AAAI10/paper/view/1879/2201.

Carr, Nicholas. 2010. *The Shallows: What the Internet Is Doing to Our Brains*. New York: W. W. Norton.

Castells, Manuel. 1996. *The Rise of the Network Society: The Information Age: Economy, Society and Culture*, vol. 1. New York: Wiley-Blackwell.

Cayley, John. 2002. "The Code is Not the Text (Unless It Is the Text)." *Electronic Book Review*. September 10. http://www.electronicbookreview.com/thread/electropoetics/literal.

———. 2004. "Literal Art: Neither Lines nor Pixels but Letters." In *First Person: New Media as Story, Performance, and Game*, edited by Noah Wardrip-Fruin and Pat Harrigan. Cambridge, MA: MIT Press. 208–17.

Christen, Kimberly. 2008. "Working Together: Archival Challenges and Digital Solutions in Aboriginal Australia." *SA Archeological Record* 8.2 (March): 21–24.

———. 2009a. "Access and Accountability: The Ecology of Information Sharing in the Digital Age." *Anthropology News* 50.4 (April): 4–5.

———. 2009b. Presentation at the Institute for Multimedia Literacy, University of Southern California, July 23.

———. N.d. *Mukurtu: Wumpurrarini-kari archive*. http://www.mukurtuarchive.org/.

Chun, Wendy Hui Kyong. 2008. *Control and Freedom: Power and Paranoia in the Age of Fiber Optics*. Cambridge, MA: MIT Press.

———. 2011. *Programmed Visions: Software and Memory*. Cambridge, MA: MIT Press.

Clark, Andy. 2004. *Natural-Born Cyborgs: Minds, Technologies, and the Future of Human Intelligence*. London: Oxford University Press.

———. 2008. *Supersizing the Mind: Embodiment, Action, and Cognitive Extension*. London: Oxford University Press.

Clement, Tanya E. 2008a. Interview with N. Katherine Hayles. Lincoln, NE, October 11.

———. 2008b. "'A Thing Not Beginning and Not Ending': Using Digital Tools to Distant-Read Gertrude Stein's *The Making of Americans*." *Literary and Linguistic Computing* 23.3 (September 16): 361–81.

Clergy of the Church of England Database. N.d. http://www.theclergydatabase.org.uk/index.html. Accessed January 30.

Cohen, Margaret. 2009. "Narratology in the Archive of Literature: Literary Studies' Return to the Archive." *Representations* 108.1 (Fall): 51–75.

Cohen, Matthew. 2011. "Design and Politics in Electronic American Literary Archives." In *The American Literature Scholar in the Digital Age*, edited by Amy Earhart and Andrew Jewell. Ann Arbor: University of Michigan Press. 228–49.

The "Colorado" Code. 1918. No. 2. NG. General Headquarters, American Expeditionary Forces, United States Army.

Connolly, Thomas, and Carolyn Begg. 2002. *Database Systems: A Practical Approach to Design, Implementation, and Management*. 3rd ed. New York: Addison-Welsey.

Crane, Gregory. 2008a. "Digital Humanities." Workshop, University of Nebraska, Lincoln, NE, October 10.

———. 2008b. Interview with N. Katherine Hayles. Lincoln, NE, October 11.

Crane, Mary Thomas. 2009. "Surface, Depth, and the Spatial Imaginary: A Cognitive Reading of *The Political Unconscious*." *Representations* 108.1 (Fall): 76–97.

Crary, Jonathan. 2001. *Suspensions of Perception: Attention, Spectacle, and Modern Culture*. Cambridge, MA: MIT Press.

Culler, Jonathan. 2010. "The Closeness of Close Reading." *ADE Bulletin* 149: 20–25.

Damasio, Antonio R. 2005. *Descartes' Error: Emotion, Reason, and the Human Brain*. New York: Penguin.

Daniel, Sharon. 2007. *Public Secrets*. *Vectors* 2.2. http://vectors.usc.edu/index.php?page=7&projectId=57.

———. 2008. Interview with N. Katherine Hayles. Santa Cruz, CA, and Hillsborough, NC, October 13.

———. 2011. *Blood Sugar*. http://bloodsugararchives.net/.

Danielewski, Mark Z. 2000. *House of Leaves*. New York: Pantheon.

———. 2007a. "*LAist* Interview: Mark Z. Danielewski." By Callie Miller. *LAist* (October 23). http://laist.com/2007/10/23/laist_interview_55.php.

———. 2007b. *Only Revolutions*. New York: Pantheon.

———. 2007c. "Revolutions 2: An Interview with Mark Z. Danielewski." By Kiki Benzon. *Electronic Book Review*. March 2. http://www.electronicbookreview.com/thread/wuc/regulated.

———. 2007d. "Spoiler Poster." *Revue Inculte* 14 (October 29).

Dannenberg, Hilary. 2008. *Coincidence and Counterfactuality: Plotting Time and Space in Narrative Fiction*. Lincoln: University of Nebraska Press.

Darabont, Frank. 2005. *Walpuski's Typewriter*. Baltimore: Cemetery Dance Publications.

Davidson, Cathy N. 2008. "Humanities 2.0: Promise, Perils, Predictions." *PMLA* 123.3 (May): 707–17.

Davidson, Cathy N., and David Theo Goldberg. 2004. "Engaging the Humanities." *Profession 2004*, 42–62.

Deacon, Terrence W. 1998. *The Symbolic Species: The Co-evolution of Language and the Brain*. New York: W. W. Norton.

Dehaene, Stanislas. 2009. *Reading in the Brain: The Science and Evolution of a Human Invention*. New York: Viking.

Deleuze, Gilles, and Félix Guattari. 1987. *A Thousand Plateaus: Capitalism and Schizophrenia*. Translated by Brian Massumi. Minneapolis: University of Minnesota Press.

DeLillo, Don. *Libra*. New York: Penguin, 1991.

Denard, Hugh. 2002. "Virtuality and Performativity: Recreating Rome's Theater of Pompey." *PAJ: A Journal of Performance and Art* 24.1 (January): 25–43.

Dennett, Daniel C. 1991. *Consciousness Explained*. New York: Back Bay Books.

DeStefano, Diana, and Jo-Anne LeFevre. 2007. "Cognitive Load in Hypertext Reading: A Review." *Computers in Human Behavior* 23.3 (May): 1616–41.

Dijksterhuis, Ap, Henk Aarts, and Pamela K. Smith. 2005. "The Power of the Subliminal: On Subliminal Persuasion and Other Potential Applications." In *The New Unconscious*, edited by Ran R. Hassin, James S. Uleman, and John A. Bargh. Oxford: Oxford University Press. 77–106.

Direct Service Guide Book and Telegraphic Cipher. 1939. NG. New York: Dun and Bradstreet, Mercantile Agency.

Dodwell, Robert, and George Ager. 1874. *The Social Code*. FB. London: Oriental Telegram Agency.

Donald, Merlin. 2002. *A Mind So Rare: The Evolution of Human Consciousness*. New York: W. W. Norton.

Doyle, Arthur Conan. (1908) 1993. "The Adventure of the Bruce-Partington Plans." In *The Case Book of Sherlock Holmes*. Hertfordshire, UK: Wordsworth Editions. 765–82.

Drucker, Johanna. 2009. *SpecLab: Digital Aesthetics and Projects in Speculative Computing*. Chicago: University of Chicago Press.

Egan, Greg. 1995. *Permutation City*. Eindhoven, Netherlands: Eos.

Emmott, Catherine. 1997. *Narrative Comprehension: A Discourse Perspective*. New York: Oxford University Press.

Ethington, Philip J. 2000. "Los Angeles and the Problem of Urban Historical Knowledge." Multimedia essay to accompany the December issue of *American Historical Review*. http://www.usc.edu/dept/LAS/history/historylab/LAPUHK.

———. 2007. "Placing the Past: 'Groundwork' for a Spatial Theory of History." *Rethinking History* 11.4 (December): 465–93.

———. 2009. Interview with N. Katherine Hayles. Los Angeles, CA, and Hillsborough, NC, January 8.

Etzioni, Oren. 2007. *Machine Reading: Papers from the AAAI Spring Symposium*. Technical report SS0-07-06.

Etzioni, Oren, Michele Banko, and Michael J. Cafarella. 2006. "Machine Reading." http://turing.cs.washington.edu/papers/aaai06.pdf.

Folsom, Ed. 2007. "Database as Genre: The Epic Transformation of Archives." *PMLA* 122.5 (October): 1571–79.

Foster, Stephen. "O Susanna!" (N.d.) 1920. In *Stephen Collins Foster: A Biography of America's Folk-Song Composer*, by Harold Vincent Milligan. New York: G. Schirmer. http://www.archive.org/stream/stephencollinsfo00millrich/stephencollinsfo00millrich_djvu.txt.

Foster, Thomas. 2005. *The Souls of Cyberfolk: Posthumanism as Vernacular Theory*. Minneapolis: University of Minnesota Press.

Francese, Joseph. 1997. *Narrating Postmodern Time and Space*. Albany: State University of New York Press.

Frank, Joseph. 1991. "Spatial Form: Some Further Reflections." In *The Idea of Spatial Form*. New Brunswick, NJ: Rutgers University Press. 107–32.

Frank, Zephyr. 2010. "Terrain of History." The Spatial History Project, Stanford University. http://www.stanford.edu/group/spatialhistory/cgi-bin/site/project.php?id=999.

Fredkin, Edward. 2007. "Informatics and Information Processing versus Mathematics and Physics." Presentation at the Institute for Creative Technologies, Marina Del Rey, CA, May 25.

Frischer, Bernard. Rome Reborn: A Digital Model of Ancient Rome. http://www.romereborn.virginia.edu.

Fukuyama, Francis. 2006. *The End of History and the Last Man*. New York: Free Press.
Fuller, Matthew. 2007. *Media Ecologies: Materialist Energies in Art and Technoculture*. Cambridge, MA: MIT Press
Gallop, Jane. 2009. "Close Reading in 2009." *ADE Bulletin* 149: 15–19.
Galloway, Alexander R. 2004. *Protocol: How Control Exists after Decentralization*. Cambridge, MA: MIT Press.
Galloway, Alexander R., and Eugene Thacker. 2007. *The Exploit: A Theory of Networks*. Minneapolis: University of Minnesota Press.
Gambrell, Alice, with Raegan Kelly. 2005. "The Stolen Time Archive," *Vectors* 1.1. http://www.vectorsjournal.net/index.php?page=7&projectId=10.
———. 2008. Interview with N. Katherine Hayles. Los Angeles, CA, and Hillsborough, NC, October 23.
General Cipher Code. 1930. New York: Railway Express Agency.
Genette, Gérard. 1983. *Narrative Discourse: An Essay on Method*. Translated by Jane E. Lewin. Ithaca, NY: Cornell University Press.
Gilmore, Paul. 2002. "The Telegraph in Black and White." *ELH* 69.3 (Fall): 805–33.
Gitelman, Lisa. 1999. *Scripts, Grooves, and Writing Machines: Representing Technology in the Edison Era*. Stanford, CA: Stanford University Press.
Glazier, Loss. 2008. *Dig[iT]al Poet(I)cs: The Making of E-Poetry*. Tuscaloosa: University of Alabama Press.
Gleick, James. 1993. *Genius: The Life and Times of Richard Feynman*. New York: Vintage.
Goodchild, Michael. 2008. "Combining Space and Time: New Potential for Temporal GIS." In *Placing History: How Maps, Spatial Data, and GIS Are Changing Historical Scholarship*, edited by Anne Kelly Knowles and Amy Hillier. Redlands, CA: ESRI. 179–198.
Greenfield, Patricia M. 2009. "Technology and Informal Education: What Is Taught, What Is Learned." *Science* 323.5910 (January): 69–71.
Griswold, Wendy, Terry McDonnell, and Nathan Wright. 2005. "Reading and the Reading Class in the Twenty-First Century." *Annual Review of Sociology* 31: 127–42.
Gromola, Diane, and Jay David Bolter. 2003. *Windows and Mirrors: Interaction Design, Digital Art, and the Myth of Transparency*. Cambridge, MA: MIT Press.
Guillory, John. 2008. "How Scholars Read." *ADE Bulletin* 146: 8–17.
———. 2010a. "Close Reading: Prologue and Epilogue." *ADE Bulletin* 149: 8–14.
———. 2010b. "Genesis of the Media Concept." *Critical Inquiry* 36 (Winter): 321–62.
Hall, Gary. 2008. *Digitize This Book! The Politics of New Media, or Why We Need Open Access Now*. Minneapolis: University of Minnesota Press.
Hall, Steven. 2007a. "The Aquarium Fragment (Prologue)." http://rawshark.ca/fragment.pdf.
———. 2007b. "Crypto-Forensics: What Are Raw Shark Negatives?" http://forums.steven-hall.org/yaf_postsm54_What-are-Raw-Shark-Texts-Negatives.aspx.
———. 2007c. Letter 175 (Negative 8). http://forums.steven-hall.org/yaf_postst97_Negative-8-Letter-175.aspx.
———. 2007d. *The Raw Shark Texts: A Novel*. Toronto: HarperPerennial.
———. (2007) 2008a. *The Raw Shark Texts: A Novel*. New York: Canongate US.
———. 2008b. "The Red Cabinet." http://forums.steven-hall.org/yaf_topics4_The-Red-Cabinet.aspx.
———. 2009. "A Progressive Documentation of Everything." http://forums.steven-hall.org/yaf_postst136_A-Progressive-Documentation-of-Everything-working-title.aspx.
Hallner, Andrew. 1912. *The Scientific Dial Primer: Containing Universal Code Elements of Universal Language, New Base for Mathematics, Etc.* San Francisco: Sunset.

Hamilton, David P. 1990. "Publishing by—and for?—The Numbers." *Science* 250.4986 (December 7): 1331–32.
———. 1991. "Research Papers: Who's Uncited Now?" *Science* 251.4989 (January 4): 25.
Hansen, Mark B. N. 2006a. *Bodies in Code: Interfaces with Digital Media*. New York: Routledge.
———. 2006b. *New Philosophy for New Media*. Cambridge, MA: MIT Press.
———. 2011. "Print Interface to Time: *Only Revolutions* at the Crossroads of Narrative and History." In *Essays on Mark Z. Danielewski*, edited by Alison Gibbons and Joe Bray. Manchester: Manchester University Press. 179–99.
Haraway, Donna. 1988. "The Science Question in Feminism and the Privilege of Partial Perspective." *Feminist Studies* 14.3: 575–99.
Harding, Sandra. 1986. *The Science Question in Feminism*. Ithaca, NY: Cornell University Press.
Harris, Trevor M., L. Jesse Rouse, and Susan Bergeron. 2010. "The Geospatial Semantic Web, Pareto GIS, and the Humanities." In *The Spatial Humanities: GIS and the Future of Humanities Scholarship*, edited by David J. Bodenhamer, John Corrigan, and Trevor M. Harris. Newark: University of Delaware Press. 124–42.
Hayles, N. Katherine. 1993. "Constrained Constructivism: Locating Scientific Inquiry in the Theater of Representation." In *Realism and Representation: Essays on the Problem of Realism in Relation to Science, Literature, and Culture*, edited by George Levine. Madison: University of Wisconsin Press. 27–43.
———. 1999. *How We Became Posthuman: Virtual Bodies in Cybernetics, Literature, and Informatics*. Chicago: University of Chicago Press.
———. 2001. "The Invention of Copyright and the Birth of Monsters: Flickering Connectivities in Shelley Jackson's *Patchwork Girl*." *Journal of Postmodern Culture*10.2 (September). http://www.iath.virginia.edu/pmc/.
———. 2002. *Writing Machines*. Cambridge, MA: MIT Press.
———. 2003. "Translating Media: Why We Should Rethink Textuality." *Yale Journal of Criticism* 16.2: 263–90.
———. 2005. "The Dream of Information: Escape and Constraint in the Bodies of Three Fictions." In *My Mother Was a Computer: Digital Subjects and Literary Texts*. Chicago: University of Chicago Press. 62–88.
———. 2007a. "Hyper and Deep Attention: The Generational Divide in Cognitive Modes." *Profession 2007*, 187–99.
———. 2007b. "Narrative and Database: Natural Symbionts." *PMLA* 112.5 (October):1603–8.
———. 2009. "RFID: Human Agency and Meaning in Information-Intensive Environments." *Theory, Culture and Society* 26.2/3: 1–24.
Heims, Steve J. 1991. *The Cybernetics Group*. Cambridge, MA: MIT Press.
Hermetic Word Frequency Counter. N.d. http: www.hermetic.ch/wfc/wfc.htm.
Hollingshead, John. 1860. "House-Top Telegraphs." In *Odd Journeys In and Out of London*. London: Groomsbridge and Sons. 233–45.
Hui, Barbara Lok-Yin. 2010. "Narrative Networks: Mapping Literature at the Turn of the Twenty-First Century." PhD diss., University of California, Los Angeles.
Hunt, Leta, and Philip J. Ethington. 1997. "The Utility of Spatial and Temporal Organization in Digital Library Construction." *Journal of Academic Librarianship* 23: 475–83.
Hutchins, Edwin. 1996. *Cognition in the Wild*. Cambridge, MA: MIT Press.
The Ideal Code Condenser: Being a Thirteen Figure Code. N.d. London: Central Translations Institute.

International Code of Signals: American Edition. 1917. FB. Washington, DC: Government Printing Office, published by the Hydrographic Office under the authority of the Secretary of the Navy.

International Code of Signals for Visual, Sound, and Radio Communication: United States Edition. 1968. FB. Washington, DC: Government Printing Office, published for the US Naval Oceanographic Office under the authority of the Secretary of the Navy.

International Morse Code (Instructions). 1945. War Department Technical Manual TM11-459. NG. Washington, DC: Government Printing Office.

James, Henry. 1898. *In the Cage*. London: Duckworth.

Jameson, Fredric. 1981. *The Political Unconscious: Narrative as a Socially Symbolic Act*. Ithaca, NY: Cornell University Press.

Jeannerod, Marc. 2002. *Le cerveau intime*. Paris: Olide Jacob.

Johnson, Addie, and Robert W. Proctor. 2004. *Attention: Theory and Practice*. Thousand Oaks, CA: Sage.

Johnson, Barbara. 1985. "Teaching Deconstrucively." In *Writing and Reading Differently: Deconstruction and the Teaching of Composition and Literature*, edited by G. Douglas Atkins and Michael L. Johnson. Lawrence: University Press of Kansas.

Johnson, Steven. 2006. *Everything Bad Is Good for You*. New York: Riverhead Trade.

Johnston, John. 1998. *Information Multiplicity: American Fiction in the Age of Media Saturation*. Baltimore: Johns Hopkins University Press.

Kahn, David. 1967. *The Code-Breakers: The Story of Secret Writing*. New York: Scribner.

Kaplan, Caren, with Raegan Kelly. 2007. "Dead Reckoning: Aerial Perception and the Social Construction of Targets." *Vectors* 2.2. http://vectors.usc.edu/projects/index.php?project=11.

———. 2008. Interview with N. Katherine Hayles. Davis, CA, and Hillsborough, NC, October 27.

Kilgarriff, Adam. 2001. "Comparing Corpora." *International Journal of Corpus Linguistics* 6.1: 1–37.

Kirschenbaum, Matthew G. 2004. "Reading at Risk: A Response." July 21. http://otal.umd.edu/~mgk/blog/archives/000563.html.

———. 2007. "How Reading Is Being Reimagined." *Chronicle of Higher Education* 54.15 (December 7): B20.

———. 2008. *Mechanisms: New Media and the Forensic Imagination*. Cambridge, MA: MIT Press.

———. 2009. Interview with N. Katherine Hayles. College Park, MD, and Hillsborough, NC, January 9.

———. 2010. "What Is Digital Humanities and What's It Doing in English Departments?" *ADE Bulletin* 150. http://www.ade.org/bulletin/index.htm.

Kittler, Friedrich A. 1992. *Discourse Networks, 1800/1900*. Translated Michael Metteer. Stanford, CA: Stanford University Press.

———. 1999. *Gramophone, Film, Typewriter*. Translated Geoffrey Winthrop-Young. Stanford, CA: Stanford University Press.

Klingberg, Torkel. 2009. *The Overflowing Brain: Information Overload and the Limits of Working Memory*. Oxford: Oxford University Press.

Knoespel, Kenneth. 2009. Interview with N. Katherine Hayles. Georgia Institute of Technology, Atlanta GA, January 15.

Kroenke, David, and David J. Auer. 2007. *Database Concepts*. 3rd ed. New York: Prentice Hall.

Lacan, Jacques. 1966. "The Instance of the Letter in the Unconscious." In *Ecrits: The First Complete Edition in English*, translated by Bruck Fink in collaboration with Hélöise Fink and Russell Grigg. New York: W. W. Norton. 412–43.

Larson, Kevin. 2004. "The Science of Word Recognition, or How I Learned to Stop Worrying and Love the Bouma." http://www.microsoft.com/typography/ctfonts/wordrecognition.aspx.

Latour, Bruno. 1992. *We Have Never Been Modern*. Cambridge, MA: Harvard University Press.

———. 1994. "On Technical Mediation—Philosophy, Sociology, Genealogy." *Common Knowledge* 3.2: 29–64.

———. 1999. *Pandora's Hope: Essays on the Reality of Science Studies*. Cambridge, MA: Harvard University Press.

———. 2007. *Reassembling the Social: An Introduction of Actor-Network Theory*. Cambridge, MA: Harvard University Press.

Laurence, David. 2008. "Learning to Read." *ADE Bulletin* 145: 3–7.

Lefebvre, Henri. (1974) 1992. *The Production of Space*. Translated by Donald Nicholson-Smith. Hoboken, NJ: Wiley-Blackwell.

Lenoir, Timothy. 2008a. Interview with N. Katherine Hayles. Durham, NC, September 3.

———. 2008b. "Recycling the Military-Entertainment Complex with Virtual Peace." http://virtualpeace.org/whitepaper.php.

Lenoir, Tim, and Eric Giannella. 2011. "Technological Platforms and the Layers of Patent Data." In *Making and Unmaking Intellectual Property: Creative Production in Legal and Cultural Perspective*, edited by Mario Biagioli, Peter Jaszi, and Martha Woodmansee. Chicago: University of Chicago Press.

Lenoir, Timothy, et al. 2008. *Virtual Peace: Turning Swords to Ploughshares*. http://virtualpeace.org/.

Lewes, Charles. 1881. "Freaks of the Telegraph." *Blackwood's Magazine* 129: 468–78.

Libet, Benjamin, et al. 1979. "Subjective Referral of the Timing for a Conscious Sensory Experience: A Functional Role for the Somatosensory Specific Projection System in Man." *Brain* 102.1: 193–224.

Liebèr, Benjamin F. 1915. *Liebèr's Five Letter American Telegraph Code*. NG. New York: Liebèr Publishing.

Liporace, Carlo Alberto. 1929. *Tourist Telegraphic Code*. FB. Naples: Tipografia A. Trani.

Liu, Alan. 2004. *The Laws of Cool: Knowledge Work and the Culture of Information*. Chicago: University of Chicago Press.

———. 2008a. Interview with N. Katherine Hayles. Los Angeles, CA, and Hillsborough, NC, October 13.

———. 2008b. *Local Transcendence: Essays on Postmodern Historicism and the Database*. Chicago: University of Chicago Press.

———. 2008c. "Re-Doing Literary Interpretation: A Pedagogy." *Currents in Electronic Literacy*. http://currents.dwrl.utexas.edu/Spring08/Liu.

Liu, Lydia. 1995. *Translingual Practice: Literature, National Culture, and Translated Modernity—China, 1900–1937*. Stanford, CA: Stanford University Press.

Lloyd, David. 2008. Interview with N. Katherine Hayles. Los Angeles, CA, and Hillsborough, NC, October 7.

Lloyd, David, and Eric Loyer. 2006. *Mobile Figures*. *Vectors* 1.2. http://vectors.usc.edu/index.php?page=7&projectId=54.

Lovecraft, Howard. P. (1937) 1999. "The Thing on the Doorstep." In *More Annotated Lovecraft*, edited by S. T. Joshi and Peter Cannon. New York: Dell.

———. (1941) 2010. *The Case of Charles Dexter Ward*. Edited by S. T. Joshi. Tampa: University of Tampa Press.

Mackenzie, Adrian. 2002. *Transductions: Bodies and Machines at Speed*. London: Continuum.

Mahlman, Jeffrey. 1970. "The 'Floating Signifier' from Lévi-Strauss to Lacan." *Yale French Studies* 48: 10–37.

Malabou, Catherine. 2008. *What Should We Do with Our Brain?* Translated by Sebastian Rand. New York: Fordham University Press.

Mangen, Anne. 2009. *The Impact of Digital Technology on Immersive Fiction Reading: A Cognitive-Phenomenological Study*. Saarbrücker, Germany: VDM Verlag.

Manovich, Lev. 2002. *The Language of New Media*. Cambridge, MA: MIT Press.

———. 2007. "Cultural Analytics: Analysis and Visualization of Large Cultural Data Sets." http://www.manovich.net/cultural_analytics.pdf.

———. 2010. "Shaping Time." http://www.flickr.com/photos/graphicdesignmuseum/sets/72157623763165627/.

———. N.d. "Database as a Symbolic Form." http://cuma.periplurban.org/wp-content/uploads/2008/06/manovich_databaseassymbolicform.pdf.

Marino, Mark. 2006. "Critical Code Studies." *Electronic Book Review* (December 4). http: www.electronicbookreview.com/thread/electropoetics/codologn.

Massey, Doreen. 1994a. "A Global Sense of Place." In *Space, Place, and Gender*. Minneapolis: University of Minnesota Press. 146–56.

———. 1994b. "Uneven Development: Social Change and Spatial Divisions of Labour." In *Space, Place and Gender*. Minneapolis: University of Minnesota Press. 86–114.

———. 2005. *For Space*. London: Sage.

Matthews, Harry. 1994. *The Journalist: A Novel*. Boston: David R. Godine.

McCarty, Willard. 2005. *Humanities Computing*. London: Palgrave.

———. 2008. "Knowing . . . : Modeling in Literary Studies." In *A Companion to Digital Literary Studies*, edited by Susan Schreibman and Ray Siemens. Oxford: Blackwell. 391–401.

———. 2009. Interview with N. Katherine Hayles. Centre for Computing in the Humanities, King's College, London, January 23.

McCarty, Willard, et al. 2009. Interview with N. Katherine Hayles. Centre for Computing in the Humanities, King's College, London, January 23.

McCarty, Willard, and Harold Short. 2009. Interview with N. Katherine Hayles. Centre for Computing in the Humanities, King's College, London, January 23.

McHale, Brian. 2011. "*Only Revolutions*, or, the Most Typical Poem in World Literature." In *Essays on Mark Z. Danielewski*, edited by Alison Gibbons and Joe Bray. Manchester: Manchester University Press. 141–59.

McLuhan, Marshall. 1964. *Understanding Media: The Extensions of Man*. New York: Mentor.

McPherson, Tara. 2008. Interview with N. Katherine Hayles. Los Angeles, CA, and Hillsborough, NC, October 20.

McPherson, Tara, and Steve Anderson, eds. N.d. *Vectors: Journal of Culture and Technology in a Dynamic Vernacular*. http://www.vectorsjournal.org/.

Menke, Richard. 2008. *Telegraphic Realism: Victorian Fiction and Other Information Systems*. Stanford, CA: Stanford University Press.

Merrick, George Byron. 1909. *Old Times on the Upper Mississippi: The Recollections of a Steamboat Pilot from 1854 to 1863*. Cleveland: Arthur H. Clark.

Miall, David S., and Teresa Dobson. 2001. "Reading Hypertext and the Experience of Literature." *Journal of Digital Information* 2.1 (August 13). http://journals.tdl.org/jodi/article/viewArticle/35/37.

Montfort, Nick, and Ian Bogost. 2009. *Racing the Beam: The Atari Computer System*. Cambridge, MA: MIT Press.

Moravec, Hans. 1990. *Mind Children: The Future of Robot and Human Intelligence*. Cambridge, MA: Harvard University Press.

———. 2000. *Robot: Mere Machine to Transcendent Mind*. New York: Oxford University Press.

Moretti, Franco. 2000. "Conjectures on World Literature." *New Left Review* 1 (January/February): 54–68.

———. 2007. *Graphs, Maps, Trees: Abstract Models for a Literary History*. New York: Verso.

———. 2009. "Style, Inc.: Reflections on Seven Thousand Titles (British Novels, 1740–1850)." *Critical Inquiry* 36 (Autumn): 134–58.

Mostern, Ruth, and I. Johnson. 2008. "From Named Place to Naming Event: Creating Gazetteers for History." *International Journal of Geographical Information Science* 22.10: 1091–1108.

Nakamura, Lisa. 2007. *Digitizing Race: Visual Cultures of the Internet*. Minneapolis: University of Minnesota Press.

National Endowment for the Arts. 2004. *Reading at Risk: A Survey of Literary Reading in America*. Research Division Report no. 46. http://www.arts.gov/pub/readingatrisk.pdf.

———. 2007. *To Read or Not to Read: A Question of National Consequence*. Research Division Report no. 47. http://www.arts.gov/research/ToRead.pdf.

———. 2009. *Reading on the Rise: A New Chapter in American Literacy*. http://www.arts.gov/research/readingonrise.pdf.

Nicolelis, Miguel. 2011. *Beyond Boundaries: The New Neuroscience of Connecting Brains with Machines—and How It Will Change Our Lives*. New York: Henry Holt.

Niederhauser, D. S., et al. 2000. "The Influence of Cognitive Load on Learning from Hypertext." *Journal of Educational Computing Research* 23.3: 237–55.

Nielsen, Jakob. 2006. "F-Shaped Pattern for Reading Web Content." *Alertbox: Current Issues in Web Usage* (April 17). http://www.useit.com/alertbox/reading_pattern.html.

———. 2008. "How Little Do Users Read?" *Alertbox: Current Issues in Web Usage* (May 6). http://www.useit.com/altertbox/percent-text-read.html.

Otis, Laura. 2001. *Networking: Communicating with Bodies and Machines in the Nineteenth Century*. Ann Arbor: University of Michigan Press.

Parton, Nigel. 2008. "Changes in the Form of Knowledge in Social Work: From the 'Social' to the 'Informational'?" *British Journal of Social Work* 38.2: 253–69.

Pendlebury, David. 1991. Letter to the editor. *Science* 251.5000 (March 22): 1410–11.

The Pennsylvania Railroad and Long Island Rail Road Telegraphic Code. 1946. FB.

Pentland, Alex. 2008. *Honest Signals: How They Shape Our World*. Cambridge, MA: MIT Press.

Peters, John Durham. 2006. "Technology and Ideology: The Case of the Telegraph Revisited." In *Thinking with James Carey: Essays on Communications, Transportation, History*, edited by Jeremy Packer and Craig Robertson. New York: Peter Lang. 137–55.

Phillips, Walter B. 1878. *The Phillips Code*. FB. Union, NJ: National Telegraph Office.

Pickering, Andy. 1995. *The Mangle of Practice: Time, Agency, and Science*. Chicago: University of Chicago Press.

Pocket Blank Code, Number 5, Containing 2000 Five-Letter Words. 1924. FB. New York: American Code Company.

"Pictured: Electric Lollipop That Allows Blind People to 'See' Using Their Tongue." 2009. *Daily Mail* (September 2). http://www.dailymail.co.uk/sciencetech/article-1210425/Blind-people-able-using-amazing-tongue-tingling-device-bypasses-eyes.html.

Pockett, E. 2002. "On Subjective Back-Referral and How Long It Takes to Become Conscious of a Stimulus: A Reinterpretation of Libet's Data." *Consciousness and Cognition* 11.2: 144–61.

Portela, Manuel. 2011. "The Book as Computer: A Numerical and Topological Analysis of *Only Revolutions*." Unpublished MS.

Prescott, George B. 1860. *History, Theory, and Practice of the Electric Telegraph*. Boston: Ticknor and Fields.

Presner, Todd S. 2006. "*Hypermedia Berlin*: Cultural History in the Age of New Media, or 'Is There a Text in This Class?'" *Vectors* 1.2. http://www.vectorsjournal.org/projects/index.php?project=60.

———. 2007. *Mobile Modernity: Germans, Jews, Trains*. New York: Columbia University Press.

———. 2008. Interview with N. Katherine Hayles. Los Angeles, CA, and Hillsborough, NC, September 3.

———. 2009a. "Digital Geographies: Berlin in the Age of New Media." In *Spatial Turns: Space, Place, and Mobility in German Literature and Visual Culture*, edited by Jaimey Fisher and Barbara Menuel. Amsterdam: Rodopi.

———. 2009b. "*HyperCities*: Building a Web 2.0 Learning Platform." In *Teaching Literature at a Distance*, edited by Anastasia Natsina and Tukis Tagialis. New York: Continuum Books.

Presner, Todd S., et al. 2008. *HyperCities*. http://www.hypercities.com/.

Pressman, Jessica. 2009. "The Aesthetic of Bookishness in Twenty-First Century Literature." *Michigan Quarterly Review* 48.4. http://hdl.handle.net/2027/spo.act2080.0048.402.

Purdy, James P., and Joyce R. Walker. 2010. "Valuing Digital Scholarship: Exploring the Changing Realities of Intellectual Work." *Profession 2010*: 177–95.

Pynchon, Thomas. 1974. *Gravity's Rainbow*. New York: Bantam.

Rabkin, Eric S. 2006. "Audience, Purpose, and Medium: How Digital Media Extend Humanities Education." In *Teaching, Technology, Textuality: Approaches to New Media*, edited by Michael Hanrahan and Deborah L. Madsen. London: Palgrave Macmillan. 135–47.

Raley, Rita. 2003. "Machine Translation and Global English." *Yale Journal of Criticism* 16.2: 291–313.

———. 2006. "Code.surface || Code.depth." *Dichtung-digital*. http://www.brown.edu/Research/dichtung-digital/2006/1-Raley.htm.

———. 2008. Interview with N. Katherine Hayles. Santa Barbara, CA, and Hillsborough, NC, October 20.

———. 2009. *Tactical Media*. Minneapolis: University of Minnesota Press.

Ramsay, Stephen. 2008a. "Algorithmic Criticism." In *A Companion to Digital Literary Studies*, edited by Susan Schreibman and Ray Siemens. Oxford: Blackwell. 477–91. Also available at http://digitalhumanities.buffalo.edu/docs/algorithmic_criticism.pdf.

———. 2008b. Interview with N. Katherine Hayles. Lincoln, NE, October 11.

Richardson, Brian. 1997. *Unlikely Stories: Causality and the Nature of Modern Narrative*. Newark: University of Delaware Press.

Ricoeur, Paul. 1990. *Time and Narrative*. Vol. 1. Translated by Kathleen McLaughlin. Chicago: University of Chicago Press.

Riddell, Allen Beye. 2011. "The Demography of Literary Forms: Extended Moretti's Graphs." Unpublished MS.

Riddell, Allen Beye, and N. Katherine Hayles. 2010. *Only Revolutions* Commentary. http://onlyrevolutions.info/.

Roberts, D. F., U. G. Foehr, and V. Rideout. 2005. *Generation M: Media in the Lives of 8–18 Year-Olds*. Kaiser Family Foundation study. http://www.kff.org/entmedia/entmedia030905pkg.cfm.

Robertson, Margaret, Andrew Fluck, and Ivan Webb. N.d. "Children, On-Line Learning and Authentic Teaching Skills in Primary Education." http://www.educ.utas.edu.au/users/ilwebb/Research/index.htm.

Rockwell, Geoffrey. 2007. TAPoR (Text Analysis Portal for Research). http://portal.tapor.ca/portal/portal.

Ross, Nelson E. 1927. *How to Write Telegrams Properly*. Little Blue Book no. 459. Edited by E. Haldeman-Julius. FB. Girard, KS: Haldeman-Julius.

Sacks, Oliver. 1998. *The Man Who Mistook His Wife For a Hat*. New York: Touchstone.

Sanders, Laura. 2009. "Trawling the Brain." *Science News* 176.13 (December 19): 16.

Saussure, Ferdinand de. (1916) 1983. *Course in General Linguistics*. Translated by Roy Harris. London: Duckworth.

Schnapp, Jeffrey. 2008. "Animating the Archive." *First Monday* 13.8 (August). http://www.uic.edu/htbin/cgiwrap/bin/ojs/index.php/fm/article/view/2218/2020.

———. 2009. Interview with N. Katherine Hayles. Palo Alto, CA, and Hillsborough, NC, January 7.

Schnapp, Jeffrey, and Todd Presner. 2009. "The Digital Humanities Manifesto 2.0." http://www.humanitiesblast.com/manifesto/Manifesto_V2.pdf.

Scholes, Robert. 1999. *The Rise and Fall of English: Reconstructing English as a Discipline*. New Haven, CT: Yale University Press.

School for Literature, Culture and Communication. Mission Statement. http://dm.lcc.gatech.edu/phd/index.php.

Scott, E. B. 1883. *The Ship Owner's Telegraphic Code*. JM. London: published by the author.

Shepperson, Alfred B. 1881. *The Standard Telegraphic Cipher Code for the Cotton Trade*. New York: Cotton Exchange Building.

Shivelbush, Wolfgang. 1987. *The Railway Journey: The Industrialization and Perception of Time and Space*. Berkeley: University of California Press.

Shnayder, Evgenia. 2009. "A Data Model for Spatial History: The Shaping the West Geodatabase." The Spatial History Project, Stanford University. http://www.stanford.edu/group/spatialhistory/cgi-bin/site/pub.php?id=23.

Siegfried, Tom. 2010. "Clash of the Quantum Titans." *Science News* 178.11 (November 20): 15–21.

Sillence, E., et al. 2007. "How Do Patients Evaluate and Make Use of Online Health Information?" *Social Science & Medicine* 64.9 (May): 1853–62.

Simondon, Gilbert. 2001. *Du mode d'existence des objets techniques*. Paris: Aubier.

Skura, Helen, Katia Nierle, and Gregory Gin. 2008. "*Romeo and Juliet*: A Facebook Tragedy." Project of the undergraduate course Literature+: Cross-Disciplinary Models of Literary Interpretation. PBWorks. http://english149-w2008.pbworks.com/Romeo%20and%20Juliet:%20A%20Facebook%20Tragedy.

Slater, Robert. 1929. *Telegraphic Code, to Ensure Secresy in the Transmission of Telegrams*. 8th ed. London: Simpkin, Marshall.

Small, Gary, and Gigi Vorgan. 2008. *iBrain: Surviving the Technological Alteration of the Modern Mind*. New York: Collins Living.

Sosnoski, James. 1999. "Hyper-Readings and Their Reading Engines." In *Passions, Pedagogies, and Twenty-First Century Technologies*, edited by Gail E. Hawisher and Cynthia L. Selfe.

Logan, UT: Utah State University Press; Urbana, IL: National Council of Teachers of English. 161–77.

Spivak, Gayatri. 1993. "The Politics of Translation." In *Outside in the Teaching Machine*. New York: Methuen. 179–200.

Standard Cipher Code of the American Railway Association. 1906. NG. New York: Committee on Standard Cipher of the American Railway Association.

Standage, Tom. 2007. *The Victorian Internet: The Remarkable Story of the Telegraph and the Nineteenth Century's On-Line Pioneers*. New York: Walker.

Stevens, C. Amory. 1884. *Watkins' Universal Shipping Code*. FB. New York.

Stewart, Garrett. 1990. *Reading Voices: Literature and the Phonotext*. Berkeley: University of California Press.

Stiegler, Bernard. 1998. *Technics and Time, 1: The Fault of Epimetheus*. Translated by Richard Beardsworth and George Collins. Stanford, CA: Stanford University Press.

———. 2009. *Technics and Time, 2: Disorientation*. Translated by Stephen Barker. Stanford, CA: Stanford University Press.

———. 2010. *Taking Care of Youth and the Generations*. Translated by Stephen Barker. Stanford, CA: Stanford University Press.

Strickland, Stephanie. 2002. *V: WaveSon.nets / Losing L'una*. New York: Penguin.

Strickland, Stephanie, and Cynthia Lawson. 2002. *V: Vniverse*. http://vniverse.com.

Summit on Digital Tools for the Humanities. 2005. *Summit on Digital Tools for the Humanities: Report on Summit Accomplishments*. Charlottesville: University of Virginia.

Svensson, Patrik. 2009. "Humanities Computing as Digital Humanities." *Digital Humanities Quarterly* 3.3 (Summer): 1–16.

Swanson, Don R., and N. R. Smalheiser. 1994. "Assessing a Gap in the Biomedical Literature: Magnesium Deficiency and Neurologic Disease." *Neuroscience Research Communication* 15.1: 1–9.

———. 1997. "An Interactive System for Finding Complementary Literatures: A Stimulus to Scientific Discovery." *Artificial Intelligence* 91.2 (April): 183–203.

Tegmark, Max. N.d. The Universes of Max Tegmark. http://space.mit.edu/home/tegmark/crazy.html.

Telegraphic Code Prepared for the Use of Officers and Men of the Navy and Their Families. N.d. [ca. 1897]. NCM. Women's Army and Navy League.

Thompson, Robert L. 1947. *Wiring a Continent: The History of the Telegraph Industry in the United States, 1832–1866*. Princeton, NJ: Princeton University Press.

Thrift, Nigel. 1996. *Spatial Formations*. London: Sage.

———. 2005. "Remembering the Technological Unconscious by Foregrounding Knowledges of Position." In *Knowing Capitalism*. London: Sage. 212–26.

Tinsley, Ron, and Kimberly Lebak. 2009. "Expanding the Zone of Reflective Capacity: Taking Separate Journeys Together." *Networks* 11.2. http://journals.library.wisc.edu/index.php/networks/article/view/190/211.

Tomasula, Steve. 2009. *TOC: A New-Media Novel*. DVD. Tuscaloosa: University of Alabama Press.

Tuan, Yi-Fu. 1977. *Space and Place: The Perspective of Experience*. Minneapolis: University of Minnesota Press.

Tukey, John W. 1997. *Exploratory Data Analysis*. Reading, MA: Addison-Wesley.

Turner, Mark. 1998. *The Literary Mind: The Origins of Thought and Language*. New York: Oxford University Press.

Unsworth, John. 2002. "What Is Humanities Computing and What Is Not?" *Jarbuch für Computerphilologie* 4: 71–83.

———. 2003. "The Humanist: 'Dances with Wolves' or 'Bowls Alone'?" Paper presented at the Association of Research Libraries conference Scholarly Tribes and Tribulations: How Tradition and Technology Are Driving Disciplinary Change, Washington, DC. http://www.arl.org/bm~doc/unsworth.pdf.

Van Alstine, H. M. 1911. *The Peace Officers' Telegraph Code Book*. FB. San Francisco: Peace Officers' Telegraphic Code Co.

Vesna, Victoria. 2007. *Database Aesthetics: Art in the Age of Information Overflow*. Minneapolis: University of Minnesota Press.

Vonnegut, Kurt. (1961) 1998. "Harrison Bergeron." In *Welcome to the Monkey House*. New York: Dell. 7–14.

Vygotsky, L. S. 1978. *Mind in Society: The Development of Higher Psychological Processes*. 14th ed. Edited by Michael Cole et al. Cambridge, MA: Harvard University Press.

Wardrip-Fruin, Noah. 2008. "Reading Digital Literature: Surface, Data, Interaction, and Expressive Processing." In *A Companion to Digital Literary Studies*, edited by Susan Schreibman and Ray Siemens. Oxford: Blackwell.

Weaver, Warren. 1949. "Memorandum." *MT News International* 22 (July): 5–6, 15. Available online at http://www.hutchinsweb.me.uk/MINI-22-1999.pdf.

———. 1955a. "Foreword: The New Tower." In *Machine Translation of Languages*, edited by William N. Locke and A. Donald Booth. Cambridge, MA: MIT Press. v–vii.

———. 1955b. "Translation." In *Machine Translation of Languages*, edited by William N. Locke and A. Donald Booth. Cambridge, MA: MIT Press. 15–23.

White, Richard. 2010. "What Is Spatial History?" The Spatial History Project, Stanford University. http://www.stanford.edu/group/spatialhistory/cgi-bin/site/pub.php?id=29.

Wiener, Norbert. 1954. *The Human Use of Human Beings: Cybernetics and Society*. New York: Da Capo.

Wilson, Timothy D. 2002. *Strangers to Ourselves: Discovering the Adaptive Unconscious*. Cambridge, MA: Harvard University Press.

Wolf, Maryanne. 2007. *Proust and the Squid: The Story and Science of the Reading Brain*. New York: Harper Perennial.

Zhu, Erping. 1999. "Hypermedia Interface Design: The Effects of Number of Links and Granularity of Nodes." *Journal of Educational Multimedia and Hypermedia* 8.3. 331–58.

Index